MATHS IN ACTION

S4[1]

Members of the
Mathematics in Action Group
associated with this book:

D. Brown
M. Brown
R.D. Howat
G. Meikle
E.C.K. Mullan
K. Nisbet

Published in 2005 by:
Nelson Thornes Ltd
Delta Place
27 Bath Road
CHELTENHAM
GL53 7TH
United Kingdom

05 06 07 08 09 / 10 9 8 7 6 5 4 3 2 1

A catalogue record of this book is available from the British Library

ISBN 0 7487 9047 0

Illustrations by Oxford Designers and Illustrators
Page make-up by Tech-Set Ltd

Printed and bound in Scotland by Scotprint

Acknowledgements

Jean Ryder (DHD Multimedia Gallery): 1; Stockbyte 35 (NT): 3; Ben Curtis/Press Association: 20; Photodisc 31 (NT): 22; Photodisc 51 (NT): 22; Digital Vision 12 (NT): 38; Photodisc 50 (NT): 48; Digital Vision 12 (NT): 52; Photodisc 10 (NT): 56, 193; Photodisc 51 (NT): 72; Photodisc 41 (NT): 77; Corel 407 (NT): 78; Peter Adams/Digital Vision BP (NT): 79, 198; Photodisc 51 (NT): 85; Columbus Foundation: 86; Photodisc 10 (NT): 96; Corel 449 (NT): 98; Digital Vision 11 (NT): 100; Photodisc 51 (NT): 105; Corel 671 (NT): 114; Corel 465 (NT): 118; Corel 760 (NT): 126; Corel 739 (NT): 128; Corel 62 (NT): 134; Corel 760 (NT): 134; Ingram ILG V2 CD4 (NT): 165; Corel 772 (NT): 214; Gerry Ellis (NT): 225.

The publishers have made every effort to contact copyright holders but apologise if any have been overlooked.

Contents

Introduction

This book has been specifically written to address the needs of the candidate following the Standard Grade Mathematics course at Foundation level. It is the second part of the course, completing the syllabus and complementing the coverage in Book S3[1].

The content has been organised to ensure that the running order of the topics is consistent with companion volumes S4[2] and S4[3] aimed at General and Credit level candidates respectively. This will permit flexibility of dual use and facilitate changing sections during the course if necessary. At places, for the purposes of practicality, some topics have been split into two parts.

The running order also permits a student attempting the National Qualifications Access 3 Cluster of Units to follow the order imposed by the descriptors of the Units. By Chapter 8 of Book S3[1], all of the required attainment targets will have been met for the Unit 'Using Mathematics 1' and the Unit test can be attempted. 'Using Mathematics 2' will be covered when the first four chapters of Book S4[1] have been completed. Further details are given on the Teacher Resource CD-ROM for this book.

Following the same model as in S3[1], each chapter follows a similar structure.
- A review section at the start ensures the knowledge required for the rest of the chapter has been revised.
- Necessary learning outcomes are demonstrated and exercises are provided to consolidate the new knowledge and skills. The ideas are developed and further exercises provide an opportunity to integrate knowledge and skills in various problem solving contexts.
- Challenges, brainstormers and investigations appear throughout to provide an opportunity for some investigative work for the more curious.
- Each chapter ends with a recap of the learning outcomes and a revision exercise which tests whether or not the required knowledge and skills addressed by the chapter have been picked up.

The final two chapters in the book are devoted to revision.
- Chapter 13 contains revision for each chapter, revisiting each of the twelve chapters in this book. This complements the equivalent chapter in Book S3[1], together providing a topic by topic revision of the whole course.
- Chapter 14 provides an opportunity to prepare for assessment. The revision exercises here give the student a chance to select strategies. In this chapter the student now encounters mixed revision in the form of six two-part tests. Part 1 of each test contains non-calculator and calculator neutral questions; part 2 provides questions where a calculator may be used.

A Teacher Resource CD-ROM provides additional material such as further practice and homework exercises and a preparation for assessment exercise for each chapter.

To assist with final revision, in association with Chapter 13 a revision checklist is included, and in association with Chapter 14 there is a 'prelim exam' laid out in a similar fashion to the final exam. A marking scheme is also included. Answers to all the questions, including those in the student book, are also provided.

1 Earnings

In 1799 Prime Minister William Pitt the Younger introduced income tax to pay for the war against Napoleon.

In 2003–04 the average wage was £476 per week. The average credit card debt was £1100 for each adult in the UK.

1 Review

◄◄ Exercise 1.1

1 In one year how many
   a months   b weeks are there?

2 How many hours are there in a typical working
   a day   b week?

3 Write these 24-hour clock times as am/pm times.
   a 15 00   b 06 00   c 08 30   d 20 45

4 How long is it from:
   a 09 00 to 13 00   b 13 30 to 17 00
   c 8.15 am to 12.30 pm   d 2.45 pm to 6 pm?

5 Write each of the following in hours:
   a 2 hours 30 minutes   (Hint: 30 minutes = 0·5 hour)
   b 5 hours 15 minutes   (Hint: 15 minutes = 0·25 hour)
   c 8 hours 45 minutes   (Hint: 45 minutes = 0·75 hour)

6 Round:
   a 8·5p to the nearest whole penny
   b 68p to the nearest 10p
   c £4·29 to the nearest £1

7 Calculate:

a 10% of £50·00   **b** 10% of £378

c 1% of £82·00   **d** 1% of £6800

e 5% of £40·00   **f** 5% of £4

g 2% of £6·00   **h** 2% of £60

8 These calculators are displaying amounts of money.
Write down the amounts shown on each in pounds and pence.

a `6·09`   **b** `8·2`   **c** `30·1`   **d** `5040·03`

9 Calculate:

a £7·25 + 89p   **b** £62 + £93·74   **c** £2780 + £189·27 + £48

d £8 − 75p   **e** £40 − 37p   **f** £239·46 − £62·87

g £7·30 × 35   **h** £364·25 × 52   **i** £2037·35 × 12

j £15 330 ÷ 12   **k** £17 992 ÷ 52   **l** £14 274 ÷ 52

10 Mr Forest earns £17 584 in a year. Mrs Forest earns £19 038.

a Calculate their total earnings.

b How much more does Mrs Forest earn than her husband?

---

2 Money calculations

Look for easy ways to do calculations.

Example 1   £7·25 + £8·47 + £2·75
= £7·25 + £2·75 + £8·47 = £10 + £8·47 = £18·47

Example 2   £52·75 − £19·99
= £52·75 − £20 + 1p = £32·76

Example 3   £42·50 × 10
= £425·00 (to multiply by 10, move the point one place to the right)

Example 4   £28 630 ÷ 100
= £286·30 (to divide by 100, move the point two places to the left)

Example 5   Estimate: **a** £52 + £87   **b** £6·85 − £2·12
c £8·95 × 4   **d** £7·88 ÷ 2

a £52 ≈ £50 and £87 ≈ £90, so estimate is £50 + £90 = £140
b £6·85 ≈ £7 and £2·12 ≈ £2, so estimate is £7 − £2 = £5
c £8·95 ≈ £9, so estimate is £9 × 4 = £36
d £7·88 ≈ £8, so estimate is £8 ÷ 2 = £4

Exercise 2.1

1 Calculate:

   a £1·65 + £1·99   **b** £6·49 + £2·49   **c** £3·15 + £2·87 + £1·85

   d £5·68 + £4·67 + £5·33   **e** £15 − £3·99   **f** £27 − £18·49

2 Calculate:

   a 28p × 10   **b** £2·77 × 10   **c** £40·50 × 100   **d** £194·35 × 100

   e £9 ÷ 10   **f** £17 ÷ 10   **g** £4629 ÷ 100   **h** £65 860 ÷ 100

3 Gordon buys a sports shirt for £26·99, shorts for £8·99 and trainers for £32·99.
How much does he spend altogether?

4 Charlotte buys a CD player for £43·49. She pays with three £20 notes.
How much change should she receive?

5 Sean buys four dining chairs. Each one costs £24·85.
What is the total cost of the chairs?

6 A monthly raffle at South Peak Sports Club has a prize of £5040.
How much does each winner receive when there are:

   a 2   **b** 3   **c** 4   **d** 5   **e** 6   **f** 7 winners?

7 Round these amounts to the nearest penny:

   a 34·1p   **b** 87·9p   **c** 38·5p   **d** £1·489

8 Round these amounts to the nearest pound:

   a £2·18   **b** £8·98   **c** £12·09   **d** £36·70

9 Round these amounts to the nearest 10 pence:

   a 49p   **b** 82p   **c** £6·85   **d** £3·39

10 Estimate:

   a £49 + £39   **b** £91 − £29   **c** £19 × 3   **d** £81 ÷ 8

   e £3·19 + £2·89 **f** £8·92 − £1·14 **g** £19·97 × 5   **h** £9·99 ÷ 5

Brainstormer

Find a quick method to do this calculation: 7 × £24·68 + 3 × £24·68

3 Wages and salaries

Example 1   Michelle works in a bank.
She earns a monthly salary of £1850.
Calculate how much she earns in a year.

Monthly salary = £1850
There are 12 months in a year.
Total earned in 1 year = £1850 × 12 = £22 200

Example 2   Simon is a car mechanic.
His total earnings in 1 year are £16 640.
He receives the same wage each week.
Calculate his weekly earnings.

Total earned in 1 year = £16 640
There are 52 weeks in a year.
Weekly wage = £16 640 ÷ 52 = £320

Exercise 3.1

 1 Neil earns £52 a day gardening.
How much is he paid for 5 days' work?

2 Jani works mornings in a shop.
She is paid £22 for each morning.
She works from Monday to Saturday.
How much does she earn in a week?

3 Sita is employed for 3 months in a travel agency.
She earns a total of £1350.
How much is she paid for each month's work?

4 Liam earns £4270 for 10 weeks' work in an
advertising office.
Calculate his weekly wage.

5 Jade works for an electronics company.
Her monthly salary is £2675.
She has been with the company for 6 months.
How much has she earned?

6 Terry earns £368 per week working as a chef.
How much does he earn in one year?

7 Sharon gets the job.
Calculate her monthly salary.

> **Accountant Required**
> £27 900 per year

8 Mandy works part-time at a bank call centre.
   Her total pay for one year is £4550.
   Calculate her weekly pay.

9 Ali applies for the post.
   If he is successful how much will he earn in a year?

Example 3 Josie works for 4·5 hours a day for 5 days.
Her hourly rate is £7·34.

    a How many hours does she work in a week?

    b What are her total weekly earnings?

    a Time worked = 4·5 × 5 = 22·5 hours

    b Total pay = £7·34 × 22·5 = £165·15

Exercise 3.2

1 On Saturday Sam works for 4 hours at a supermarket.
   The rate is £4·75 per hour. How much does he earn?

2 John earns £5·78 an hour as a night security guard.
   What is he paid for a 10 hour shift?

3 Peter the plumber is paid £13·50 for an hour's work.
   How much is he paid for a job that lasts 3 hours?

4 Maggie works 7 hours each day from Monday to Friday as a receptionist.
   Her hourly rate is £6·20.

    a How much does she earn each day?

    b Calculate her total earnings for the week.

5 Kylie earns £7·35 for each hour she works.
   The table shows her earnings for different numbers of hours worked.
   Copy and complete the table.

Number of hours	1	2	3	4	5	6	7	8
Amount earned (£)	7·35	14·70						

6 Alan, an electrician, works a 35 hour week.
   His rate of pay is £10·65 per hour.
   Calculate his total weekly earnings.

7 Mr Taylor is employed in the textile industry.
He is paid £8·26 per hour.
How much does he earn on a day when he
works 7·5 hours?

8 Mrs Green works for 6 hours a day from Monday to Friday plus 4 hours on
Saturday. She is paid £9·48 per hour.
 a How many hours does she work in a week?
 b What are her total weekly earnings?

Investigation

Look at job advertisements in newspapers.
Find examples of wages, salaries and hourly rates of pay.

4 Time-sheets and overtime

Time-sheets

Here is an example of a time-sheet.

TIME-SHEET				
Neil Morrison		**Employee No. 1476**		**Week No. 10**
	In	**Out**	**In**	**Out**	**No. of hours worked**
Mon	08 00	12 00	13 00	17 00	8
Tue	08 00	12 00	13 30	17 30	8
Wed	08 30	12 30	13 00	16 30	7·5
Thu	08 00	13 00	14 00	17 00	8
Fri	08 00	12 30	14 00	16 30	7
				Total = 38·5 hours

Look at Monday.
  Neil arrived at 8 am and left for lunch at 12 noon = 4 hours worked
  He then came back at 1 pm and worked until 5 pm = 4 hours worked
  Neil worked 8 hours on Monday.

Exercise 4.1

1 On Saturday Len starts work at 7.30 am and finishes at 1 pm.
How many hours does he work on Saturday?

2 On Tuesday, Sue arrives at work at 9 am and goes for lunch at 1 pm.

  a How many hours has she worked?

  b She returns at 2 pm and finishes work at 5.30 pm. How many hours does she work after lunch?

  c How many hours has she worked on Tuesday?

3 Alice starts work at 8.30 am. She has one hour for lunch. She works a total of 8 hours. What time does she finish work in the afternoon?

4 On night shift Dan clocks on at 10.45 pm. He has a break from 3 am to 3.45 am. He finishes his shift at 7.15 am. How many hours does he work?

5 a Use Neil's time-sheet below to find the number of hours he worked on:
    i Monday   **ii** Tuesday   **iii** Wednesday   **iv** Thursday   **v** Friday.

  b What is his total number of hours for this week?

TIME-SHEET
Neil Morrison   Employee No. 1476   Week No. 11

	In	Out	In	Out	No. of hours worked
Mon	08 00	12 00	13 00	16 00	...
Tue	08 30	12 00	13 30	17 30	...
Wed	08 30	12 00	13 00	17 30	...
Thu	08 00	12 30	13 00	17 00	...
Fri	09 00	12 30	14 00	17 30	...
				Total	... hours

Brainstormer

Javed works a total of 37·5 hours each week (Monday to Friday).
He starts work at 8.30 am. He has a 75 minute lunch break each day.
He finishes at the same time each day. What time does he finish work?

Overtime

Any extra hours worked are called overtime.
Often, as a reward for working overtime, a higher rate is paid.
If overtime is paid at double time the rate per hour is doubled.

Example   Marie's basic rate of pay is £5·30 per hour.
When she works overtime on Sunday she is paid double time.

a How much is she paid for 1 hour of overtime?

b How much is she paid for 3 hours of overtime?

a 1 hour of overtime = £5·30 × 2 = £10·60

b 3 hours of overtime = £10·60 × 3 = £31·80

Exercise 4.2

1 Stella's basic rate of pay is £6 per hour.
She is paid double time for any overtime she works.

a How much is she paid for one hour's overtime?

b On one weekend she works 5 hours' overtime.
How much does she earn?

2 Karl's basic hourly pay is £8·50 per hour. He is paid double time for any overtime he works.

a How much is he paid for one hour's overtime?

b In one week he works 4 hours' overtime.
How much does he earn?

3 Calculate the overtime rate at double time for these basic rates.

    a £5·20   **b** £7·80

    c £6·85   **d** £8·78

4 Bernie's basic rate of pay is £9 per hour.
On Monday she works for 7 hours at the basic rate.
She also does one hour's overtime paid at double time.

a How much is she paid for the overtime?

b How much is she paid altogether for Monday's work?

5 Jake's basic rate of pay is £7·40 an hour.
On a Saturday he works for 5 hours at the basic rate.
He also does 3 hours of overtime.

a How much is he paid for the 5 hours?

b Overtime is paid at double time. How much is he paid for:
    i 1 hour of overtime
    ii 3 hours of overtime?

c What is his total pay for Saturday?

6 Mrs Scott's basic rate is £8·25 per hour.
In one week she works for 36 hours.

   a Calculate her basic wage.

   b She also does 4 hours' overtime which is paid at double time.
     How much is she paid for the overtime?

   c What are her total earnings for the week?

7 This is Mr Richardson's time-sheet for Saturday.

	In	Out	In	Out	No. of hours worked
Sat	08 00	13 00	14 00	18 00	9

His basic rate of pay is £7·90 an hour.
Any work done after 13 00 is paid at double time.

   a How much does he earn at the basic rate?

   b Calculate his overtime pay.

   c How much is he paid altogether for the Saturday?

5 Piecework

Some workers are paid a set amount for each item they
make or each task they complete. This is called piecework.

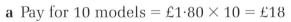

Example   Tess hand-paints model animals.
             She is paid £1·80 for each model.
             How much is she paid for painting
             a 10 models   **b** 50 models?

             a Pay for 10 models = £1·80 × 10 = £18
             b Pay for 50 models (50 = 10 × 5) = £18 × 5 = £90

Exercise 5.1

1 Graham is paid £1·50 for each car he washes.
How much is he paid for washing 8 cars?

2 Tracy charges £8 for each portrait she sketches.
How much does she earn on a day when she does
12 sketches?

3 A photographer charges £4 per pupil.
  How much does he charge for photographing 30 pupils?

4 To raise money at the school fair Mrs Flowers does face painting.
  She charges 50p for each face.
  How much money does she raise by painting 75 faces?

5 Chris is a carpenter. He receives an order to make eight chairs.
  He is paid £75 for each chair.

  a How much is he paid for the chairs?

  b He also is asked to make three tables at £145 each.
    How much is he paid for the tables?

6 Josie is paid 75p for every 10 leaflets she delivers.
  How much does she get for delivering
  a 100   **b** 300 leaflets?

7 Susie earns some extra money by typing documents for people.
  She charges 30p per 100 words.
  How much does she charge for typing
  a 600 words   **b** 1000 words   **c** 9000 words?

8 Helen has a holiday job picking fruit. She is paid £1·20 for each box she fills.
  The table shows the number of boxes she fills in her first week.
  Copy and complete the table.

Day	Mon	Tue	Wed	Thu	Fri
No. of boxes	20	25	27	28	30
Amount earned (£)					

9 Sam runs a carpet cleaning business.
  The table shows his charges.

	Charge (£)
Single carpet	20
Two bedroom house	70
Three bedroom house	85

One week he cleans:

● 8 single carpets

● 5 houses with two bedrooms

● 3 houses with three bedrooms.

How much does he earn?

Brainstormer

Eric services central-heating systems.
He works 35 hours in a week
He services 20 systems a week.
He is paid £25 for each one he services.
A company offers to employ him to do the same work on an hourly rate of £14.
Is he better off being paid a weekly wage or on piecework?

6 Commission

Some sales people are paid according to how much they sell.
This is called commission.

Example 1   Peter sells mobile phones.
        For each phone he sells he is paid a commission of £8.
        How much commission is he paid for selling 40 phones?

Commission earned = £8 × 40 = £320

Exercise 6.1

1 Angela sells carpets.
She is paid commission of £15 for each one she sells.
How much commission does she earn in a day when she sells four carpets?

2 Jenny works in an art studio.
She receives a commission of £5 on each picture she sells.
How much commission does she earn in a week when she sells 38 pictures?

3 Sheila sells satellite TV systems.
Her commission is £25 for each package sold.
The table shows her sales for one week.
Calculate her commission for each day.

Day	Mon	Tue	Wed	Thu	Fri
No. of packages	4	3	6	2	7
Commission (£)					

4 Lucy's job is to get people to join a health club.
For each new member she recruits she is paid £17·50.
How much is her commission for recruiting six new members?

5 Alan sells burglar alarm systems.
His commission is £32·50 for each sale.
How much commission does he earn for selling eight systems?

 6 Imran sells magazines. He is paid 18p for each one he sells.
How much does he earn when he sells 60 magazines?

7 Christine sells Christmas trees in a garden centre.
She is given a commission of 75p for each tree sold.
How much commission is she paid in a day when she sells 37 trees?

8 Tim's ticket agency is paid a commission of £1·50 for each ticket they sell for a concert. How much commission is earned for selling 800 tickets?

> **Often commission is calculated as a percentage of the value of the sales.**
>
> *Example 2* Mary sells bathroom suites.
> Her commission is 2% of the value of her sales.
> How much commission does she receive for selling a bathroom suite priced at £675?
>
> 1% of £675 = £675·00 ÷ 100 = £6·75
> (to divide by 100, move the point two places to the left)
> 2% of £675 = £6·75 × 2 = £13·50

Exercise 6.2

1 Kevin sells kitchen units.
He is paid a commission of 1% of the value of his sales.
How much commission is he paid when he sells kitchen units worth £900?

2 Moira works in a shoe shop.
She has a basic wage and is also paid a commission of 1% of her sales.
How much commission does she get for selling a £60 pair of shoes?

3 Sales staff at a tailor's shop earn a commission of 2%.
How much commission is earned on clothes sales of value:
  a £89   **b** £200   **c** £450?

4 Staff at Dickie's Double Glazing are paid 10% of the value of their sales.
Ernie gets a contract for £2500 worth of windows.
How much commission can he expect?

5 Ron sells school textbooks. His commission is 5% of the value of his sales.
How much commission will he receive for selling:

   a £600 of maths books

   b £1700 of science books?

6 Mrs Wood sells furniture. She is paid commission of 8% of the value of her sales.
How much does she get for these sales?

   a Dining table and chairs of value £830

   b Bedroom suite of value £1740

   c Three-piece suite of value £999

7 An estate agent receives a commission of 2% on his house sales.
Calculate the commission for selling:

   a a semi-detached house for £78 000

   b a flat for £48 500

   c a detached house for £159 900.

---

7 Deductions and take-home pay

Workers receive payslips that show their earnings.
These might include overtime, commission and bonuses.
Most workers pay income tax and National Insurance (NI).
Many workers also pay into a pension fund.

The following terms are used.
- Gross pay: total earnings including basic pay, overtime, bonuses, commission.
- Deductions: money taken to cover income tax, National Insurance and pensions.
- Take-home pay: gross pay – deductions

Example 1   These amounts are deducted from Yvonne's
weekly wage:

Income tax	£24·72
National Insurance	£8·26
Pension	£9·70

Calculate the total amount deducted.

Total deductions = £24·72 + £8·26 + £9·70 = £42·68

Example 2   Mr Sharp's monthly gross pay is £1560.
His total deductions add up to £245·90.
Calculate his take-home pay.

Take-home pay = gross pay − deductions
                 = £1560 − £245·90 = £1314·10

Exercise 7.1

1 Calculate the total deductions from Yvonne's wages for a week when she pays:

Income tax   £25·00
National Insurance   £8·00

2 Calculate the total deductions from George's earnings for these two weeks.

Week 1		Week 2	
Income tax	£42·70	Income tax	£72·58
National Insurance	£14·60	National Insurance	£19·49
Pension	£9·80	Pension	£13·54

3 This part of Tim's payslip shows his deductions.
Find the total of the deductions.

Income tax	National Insurance	Pension	Total deductions
£230·00	£115·00	£96·00	

4 Harry's gross pay for one week is £348.
His total deductions for the week add up to £97.
Calculate his take-home pay.

5 The table shows Linda's deductions.
Calculate her total deductions.

Income tax	National Insurance	Pension	Total deductions
£327·29	£167·84	£156·85	

6 Anne's payslip shows that her gross pay is £2853·60.
Her total deductions add up to £765·20.
Calculate her take-home pay.

7 This is part of Saeed's payslip.
Calculate his take-home pay for the week.

Gross pay
£342·70
Deductions
£73·95
Take-home pay

8 This column of Janice's monthly payslip
shows her gross pay and deductions.
Calculate her take-home pay.

Gross pay
£3048.04
Deductions
£1074.28
Take-home pay

Example 3

This is Mr Mason's payslip.

Calculate:

a his gross pay

b his total deductions

c his take-home pay.

Name A. Mason	Employee number 123	NI number YM23068A	Week number 12
Basic pay £653·60	Overtime £86·70	Bonus £25	Gross pay
Income tax £92·10	NI £36·60	Pension £42·40	Total deductions
			Take-home pay

Gross pay = £653·60 + £86·70 + £25 = £765·30

Total deductions = £92·10 + £36·60 + £42·40 = £171·10

Take-home pay = gross pay − total deductions = £594·20

Exercise 7.2

1 This is Mrs Baker's weekly payslip.

Name B. Baker	Employee number 14	NI number YK730571B	Week number 35
Basic pay £308·40	Overtime £35·60	Bonus –	Gross pay £344·00
Income tax £58·75	NI £26·25	Pension £20	Total deductions £105·00
			Take-home pay

a Check that the gross pay calculation is correct.

b Check that the total deductions calculation is correct.

c Calculate Mrs Baker's take-home pay.

2 This is Mr Shepherd's weekly payslip.

Name U. Shepherd	Employee number 102	NI number YM805274A	Week number 40
Basic pay £183·68	Overtime £59·04	Bonus £10·00	Gross pay
Income tax £28·57	NI £11·84	Pension £8·58	Total deductions
			Take-home pay

Calculate: **a** his gross pay

b his total deductions

c his take-home pay.

3 This is Mrs Parker's payslip for June.

Name S. Parker	Employee number 34	NI number YK537276B	Week number 3
Basic pay £1580·10	Overtime –	Bonus £63·60	Gross pay
Income tax £235·50	NI £136·70	Pension £82·30	Total deductions
			Take-home pay

Calculate: **a** her gross pay

b her total deductions

c her take-home pay.

Brainstormer

This is Mr Carpenter's weekly payslip.

Name B. Carpenter	Employee number 27	NI number YM987654A	Week number 26
Basic pay £213·65	Overtime £39·37	Bonus £****	Gross pay £277·02
Income tax £****	NI £12·49	Pension £9·36	Total deductions £****
			Take-home pay £224·92

Calculate:

a the total deductions   **b** his bonus   **c** the income tax paid.

8 Savings

Example 1   Patrick saves £12·50 each week.
How much does he save in 6 weeks?

Amount saved = £12·50 × 6 = £75

Exercise 8.1

1 Lin puts 75p into her piggy bank each week.
How much does she save in 5 weeks?

2 Each week George saves £2·50 from his pocket money.
How much has he saved in 10 weeks?

3 Jasmine saves the same amount each week. After 8 weeks she has a total of £84.
How much does she save each week?

4 After 10 months of regular saving Mr Keats saves a total of £480.
How much does he save each month?

5 Rick saves £8·60 each week. The table shows the number of weeks and the
total amount saved. Copy and complete the table.

Number of weeks	1	2	3	4	5	6	7
Total amount saved	£8·60	£17·20					

6 Ursula saves £7·75 out of her weekly wage towards her pension.
How much does she save in one year?

7 Garry saves the same amount each month for 12 months.
He has saved a total of £675. How much does he save each month?

◀◀ RECAP

Calculating with money
You should be able to add, subtract, multiply and divide amounts of money,
using quick methods where possible.

Wages and salaries
You should be able to calculate weekly wages and monthly salaries.
This may include calculations involving time-sheets, overtime (double time),
piecework and commission.

Deductions and take-home pay
You should understand and be able to do calculations involving:
gross pay, deductions and take-home pay.

> Take-home pay = gross pay − deductions

Savings
You should be able to do calculations involving regular savings.

1 Calculate:

   a 39p + 57p + 61p   **b** £3·48 + £2·55 + £4·45

   c £6·75 − 95p   **d** £17·30 − £6·99

   e £53·60 × 100   **f** £53·60 ÷ 10

2 a Harry, a caretaker, earns £260·75 a week.
   How much does he earn in 10 weeks?

  b Sharda works in a town planning department.
   In 3 months she earns a total of £8700. Calculate her monthly salary.

3 This is part of Melanie's time-sheet.

	In	Out	In	Out	No. of hours worked
Mon	08 30	13 00	14 00	17 30	

  a How long did she work on the Monday?

  b Her rate of pay is £6·40 per hour.
   How much did she earn on Monday?

4 Sandy makes pots out of clay for a craft shop.
  He is paid £12·50 for each pot.
  How much is he paid for:

  a 6

  b 10 pots?

5 a Rachel's job is selling car breakdown policies.
   She receives £7·50 for each new customer.
   How much is she paid for selling seven new policies?

  b Tariq sells advertising space in a newspaper.
   He is paid a commission of 5% of the value of his sales.
   How much commission does he earn on sales of value £800?

6 a Over one year Jill earns £14 326. How much does she earn each week?

  b Hanif's monthly salary is £2491·70.
   Calculate his total pay for one year.

7 Sue normally works a 35 hour week. Her rate of pay is £8·48 per hour.

  a Calculate her basic wage.

  b One weekend she works 3 hours' overtime at double pay.
   How much is she paid for:
    i 1 hour of overtime
    ii 3 hours of overtime?

8 This is Ruth's payslip for February.

Name R. Thatcher	Employee number 18	NI number YK865327B	Week number 11
Basic pay £1845·30	Overtime £43·96	Bonus £22·75	Gross pay
Income tax £315·50	NI £152·59	Pension £108·53	Total deductions
			Take-home pay

Calculate: **a** her gross pay
  b her total deductions
  c her take-home pay.

9 a Martha saves £15·50 each month.
  Calculate the total amount she saves in one year.

  b Melvin saves the same amount each week for 20 weeks.
  He has saved a total of £125.
  How much does he save each week?

2 Proportion

The Scottish parliament is based on Proportional Representation (PR).

PR is a system of voting designed so that the number of MPs elected from each party reflects the share of votes cast for each party.

1 Review

◀◀ Exercise 1.1

1 Calculate:

   a 16×3 **b** 37×5 **c** 182×9 **d** 295×7

   e $64 \div 4$ **f** $96 \div 8$ **g** $254 \div 2$ **h** $861 \div 7$

2 Calculate:

   a $6{\cdot}3 \times 10$ **b** $0{\cdot}52 \times 10$ **c** $5{\cdot}8 \times 100$ **d** $0{\cdot}49 \times 100$

   e $4{\cdot}1 \div 10$ **f** $62 \div 10$ **g** $63{\cdot}8 \div 100$ **h** $205 \div 100$

3 a There are 7 days in one week. How many days are there in 5 weeks?

   b There are 24 hours in one day. How many hours are there in 7 days?

   c There are 60 minutes in one hour. How many minutes are there in 10 hours?

4 a 1 mile = 1·6 km. Change 6 miles to kilometres.

   b 1 kg = 2·2 pounds. Change 8 kg to pounds.

   c 1 inch = 2·54 cm. Change 10 inches to centimetres.

5 a Two packets of biscuits cost 70p. How much does one packet cost?

   b Three cups of tea cost £1·80. How much does one cup of tea cost?

   c Five cups of coffee cost £3·50. How much does one cup of coffee cost?

6 a A model car is 4·5 cm long.
   Each centimetre stands for 1 metre.
   How long is the actual car?

   b A model train is 3 cm long.
   Each centimetre stands for 5 metres.
   How long is the actual train?

7 James takes two paces and covers 120 cm.

  a How much does he cover in one pace?

  b How far does he go in five paces?

8 Donna parks in a car park.
The graph shows the cost.

How much does it cost to park for:

  a 8 hours   **b** 4 hours

  c 2 hours   **d** 1 hour?

---

2 Rates and unit value

'per' means 'for each'. A phrase that contains the word 'per' is called a **rate**.

Example   Sean walks at 5 km per hour (km/h).
(This means that if Sean walked at this speed for one hour he would travel 5 km.)

How far would he walk in:

  a 4 hours   **b** half an hour?

  a In 4 hours he would walk $5 \times 4 = 20$ km.

  b In half an hour he would walk $5 \times \frac{1}{2} = 2{\cdot}5$ km.

Exercise 2.1

1 The entry charge for the zoo is £4 per person.
How much would it cost for three people to visit?

2 Gerry's taxi fare costs £2·50 per mile.
What is the cost of a 6 mile journey?

3 Phil's phone call costs 6p per minute.
What is the cost of his 12 minute call?

4 Diane's curtain material costs £8·50 per metre.
How much does it cost her to buy 5 metres of material?

5 Wendy runs at 10 km per hour (km/h).
How far does she run in 1·5 hours?

6 At Tina's local garage the petrol pump delivers
20 litres per minute.
It takes Tina 2·5 minutes to fill her tank.
How much petrol has she put in her tank?

7 Tony types at 30 words per minute.
How many words can he type in 25 minutes?

8 Ruth pays 83p per litre for petrol at her local garage.
How much does she pay for 40 litres?

9 Jules pays £250 per month for his flat.
What is the total cost of his rent for one year?

10 Tania's electricity costs her 6·5p per unit.
How much does it cost for 700 units?

3 Finding the rate

Example   For 5 hours' work Paul is paid £30.
Calculate his rate of pay per hour.

For 5 hours he earns £30.
For 1 hour he earns £30 ÷ 5 = £6.
So the rate of pay is £6 per hour.

Exercise 3.1

1 Debbie is paid £35 for 7 hours' work.
Calculate her rate of pay per hour.

2 An 8 line advert in the *Evening Reporter* costs £24.
Calculate the cost per line of the advert.

3 Carol's computer takes 6 minutes to print a 24 page document.
Find the rate of printing in pages per minute.

4 Hannah cycles 30 km in 2 hours.
At what rate is she cycling in kilometres per hour?

5 In 10 minutes a rowing team manage 300 strokes.
Calculate their rowing rate in strokes per minute.

6 Don uses 4 litres of varnish to cover 48 m² of floor.
What area of floor would 1 litre cover?

7 Rashid's heart beats 180 times in 3 minutes.
Find the rate that his heart beats per minute.

 8 Karen hires a rowing boat for 5 hours. It costs her £37·50.
Find the hire rate per hour.

9 Jane drives 342 km on 36 litres of petrol.
Calculate her car's fuel consumption in kilometres per litre.

10 Terry earns £29 484 in a year.
What is his monthly rate of pay?

---

4 Using the rate

Example 1   A pipeline delivers 15 litres of water in 5 seconds.
   a Calculate the rate of flow in litres per second.
   b How many litres will the pipeline deliver in 8 seconds?

   a The rate = 15 ÷ 5 = 3 litres per second
   b In 8 seconds the volume of water delivered = 3 × 8 = 24 litres

---

Exercise 4.1

 1 Elaine hires a floor sander for 4 days. It costs her £40.
   a Calculate the daily rate for hiring the sander.
   4 days cost £40
   Cost per day = £40 ÷ 4 = £...
   b How much would it cost to hire it for 5 days?
   5 days will cost 5 × £... = £....

2 Kalil takes 20 paces in 10 seconds.
   a How many paces does Kalil take in 1 second?
   b How many paces will he take in 15 seconds?

3 In an experiment a candle 12 cm tall burns for 3 hours.
   a Calculate the rate at which the candle burns in
   centimetres per hour.
   b Another candle of the same type lasts for 5 hours.
   How tall is this candle?

4 Diana uses up 30 calories cycling for 5 minutes.

  a How many calories does she use per minute when cycling?

  b How many calories will she use if she cycles for 12 minutes?

 5 John runs for 40 minutes. He reckons that he has used 440 calories.

  a How many calories does John use in 1 minute when running?

  b How many calories would he use if he ran for 60 minutes?

6 The Smiths hire a company to do a buffet for their wedding guests. The cost for 45 guests is £360.

  a Calculate the cost per guest.

  b How much would it cost if there were 50 guests?

7 Mr Taylor buys 12 m² of carpet which costs him £180.

  a Find the cost per square metre of the carpet.

  b Calculate the cost of 16 m² of the same carpet.

8 There are 82 g of fat in 100 g of butter.

  a What weight of fat is there in 1 g of butter?

  b Calculate the amount of fat in a 250 g packet of butter.

Example 2   Bella changes £10 into Swiss francs.
           She receives 23 francs.
           How many francs would she get for £200?

           £1 is worth 23 ÷ 10 = 2·3 francs
           £200 is worth 2·3 × 200 = 460 francs

Exercise 4.2

 1 Glen's library book is 3 days overdue.
    He pays a fine for each day.
    In total he pays a fine of 18p.
    Glenda's book is 5 days overdue.
    What fine should she pay?   (Hint: find the fine for 1 day.)

2 It takes Mr Keane 40 minutes to mark 8 test papers.
    How long will it take him to mark 15 test papers at the same rate?

3 Mrs Benson is paid £56 for 7 hours' work.
    How much would she earn on a day when she works for 8 hours?

4 A 5 kg bag of potatoes costs 80p.
    How much would 6 kg of potatoes cost?

5 Ian changes £10 into dollars.
   He is given $18.
   How many dollars would he get for £50?

6 Iris changes £100 into euro (€).
   She receives €145.
   How much would she get if she changed £400?

7 Sam finds that the weight of 10 ml of mercury is 136 g.
   Calculate the weight of 15 ml of mercury.

8 George's gas bill has a charge of £19·50 for 300 units.
   What would be the charge for 500 units?

Brainstormer

David is a lorry driver.
On Monday it takes him 6 hours to drive 540 km.
On Tuesday he goes 752 km in 8 hours.
Calculate his speed on each day in kilometres per hour.
On which day was his speed greater?

5 Best buys at the supermarket

Example 1   Which is the better buy?

2 Light Bulbs
for 90p

5 Light Bulbs
for £2.00

Find the cost per bulb for each offer.
90p ÷ 2 = 45p per bulb   £2 ÷ 5 = 200p ÷ 5 = 40p per bulb

The 5 bulb offer is the better buy.

Note: you will often find that it is better to change pounds to pence
when calculating the cost of one item.

Exercise 5.1

 1

| STARRY SOAP |
| 96P   3 BARS |

Starry Soap
£1·50   5 bars

    a Calculate the cost of one bar for each offer.
      Copy and complete:
      96p ÷ 3 = ... per bar; £1·50 ÷ 5 = 150p ÷ 5 = ... per bar
    b Which is the better buy?

 2

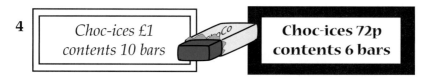

Wholemeal Flour
£1·20   2 kg

Wholemeal Flour
£1·65   3 kg

    a Calculate the cost of 1 kg for each offer.
      Copy and complete:
      £1·20 ÷ 2 = 120p ÷ 2 = ... per kg; £1·65 ÷ 3 = 165p ÷ 3 = ... per kg
    b Which is the better buy?

3

**SANDWICH BISCUITS
56P   PACK OF 7**

Sandwich Biscuits
45p   pack of 5

    a Calculate the cost per biscuit for each offer.
    b Which is the better buy?

4

*Choc-ices £1
contents 10 bars*

**Choc-ices 72p
contents 6 bars**

    a Calculate the cost of one choc-ice for each offer.
    b Which is the better buy?

5

**FRESH JUICE
£2·64   4 cartons**

**FRESH JUICE
£3·60   6 cartons**

    a Calculate the cost of one carton for each offer.
    b Which is the better buy?

6

| Christmas Turkey |
| £10·50   7 kg |

| Christmas Turkey |
| £14·40   9 kg |

a Calculate the cost per kilogram for each offer.

b Which is the better buy?

Example 2

| Top Taste Tea Bags |
| £2·88 |
| contents 160 bags |

| Top Taste Tea Bags |
| £3·84 |
| contents 240 bags |

a Calculate the cost per tea bag for each offer.

b Which is the better buy?

a Using a calculator:
   £2·88 ÷ 160 = 288 ÷ 160 = 1·8 p per bag
   £3·84 ÷ 240 = 384 ÷ 240 = 1·6p per bag

b The 240 bag offer is better.

Exercise 5.2

1

| Minty Toothpaste |
| £1·50   Contents 75 ml |

| Minty Toothpaste |
| £1·80   Contents 100 ml |

a Calculate the cost per millilitre for each tube.

b Which is the better buy?

2

| Wheat Breakfast Bars |
| 84p   contents 12 bars |

| Wheat Breakfast Bars |
| £1·20   contents 15 bars |

a Calculate the cost per bar for each offer.

b Which is the better buy?

3

| SPICY PICKLES |
| £1·50   CONTENTS 250 g |

| Spicy Pickles |
| £1·75   contents 350 g |

a Calculate the cost per gram for each jar.

b Which is the better buy?

4

Classic Coffee
£3·80   contents 200 g

Classic Coffee
£6 contents 300 g

a Calculate the cost per gram for each jar.

b Which is the better buy?

5

Sudso Soap Powder
£2·79   contents 4·5 kg

Sudso Soap Powder
£1·75   contents 2·5 kg

a Calculate the cost per kilogram for each packet.

b Which is the better buy?

6

Easy-Klean
Washing-up Liquid
80p   contents 0·5 litre

EASY-KLEAN
WASHING-UP LIQUID
£1·50 CONTENTS 1·25 LITRES

a Calculate the cost per millilitre for each container.

b Which is the better buy?

Investigation

Supermarkets usually display labels which show the price per kilogram,
per 100 grams, per litre, etc. This helps shoppers compare prices.

Find an item which is sold in different sizes. Decide which size is the best value.

It is not always the biggest container.

Repeat for other items.

6 In proportion

Example 1   Jake mixes sand and cement to make a patio.
He uses 1 part sand and 3 parts cement.

This means for every shovel of sand he must use 3 shovels of cement
or for every bag of sand he must use 3 bags of cement
or for every kilogram of sand he must use 3 kilograms of cement.

When he uses 5 kg of sand he must use 3×5 kg = 15 kg of cement.

Exercise 6.1

1 Jade mixes orange cordial with water to make orange juice.
She uses 1 part cordial with 4 parts water.
How many glasses of water should she use with:

a 1 glass of cordial

b 3 glasses of cordial?

2 Mac mixes 1 part oats with 2 parts water to make porridge.
To make porridge for his family he puts 6 cups of oats in a pan.
How many cups of water should he add?

3 Flora uses a mix of 1 part blue to 2 parts yellow to make green paint.
Copy and complete the table.

Blue	Yellow
1 ml	2 ml
2 ml	4 ml
3 ml	
	8 ml
10 ml	
	30 ml

4 Flora mixes white and black to get the shade of grey she wants.
She uses a mix of 1 part white to 5 parts black.
Copy and complete this table.

White	Black
1 ml	
2 ml	
4 ml	
	25 ml
10 ml	
	100 ml

5 Brian puts antifreeze in the radiator of his car.
He uses 1 part of antifreeze to 8 parts of water.
How much water should he use with 2 litres of antifreeze?

6 Lochside School's policy for educational trips is to send 1 teacher for every 10 pupils.

a How many pupils can go with 3 teachers?

b How many staff are needed to accompany 80 pupils?

Example 2 Barry was going to make an apple crumble.
He found this recipe for the crumble mix.

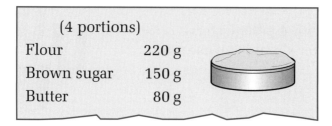

(4 portions)
Flour   220 g
Brown sugar   150 g
Butter   80 g

How much of each ingredient would he need to make a crumble for:
a 2   **b** 6 people?

a For 2 portions divide the 4 portion amounts by 2:
Flour   110 g
Brown sugar   75 g
Butter   40 g

b For 6 portions multiply the 2 portion amounts by 3:
Flour   330 g
Brown sugar   225 g
Butter   120 g

Exercise 6.2

1

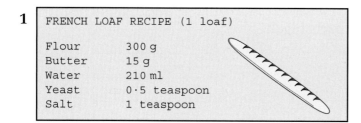

```
FRENCH LOAF RECIPE (1 loaf)

Flour      300 g
Butter     15 g
Water      210 ml
Yeast      0·5 teaspoon
Salt       1 teaspoon
```

How much of each ingredient is needed to make:
a 2   **b** 5 loaves?

2

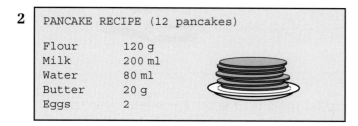

```
PANCAKE RECIPE (12 pancakes)

Flour      120 g
Milk       200 ml
Water      80 ml
Butter     20 g
Eggs       2
```

How much of each ingredient is needed to make:
a 6   **b** 18 pancakes?

3

```
FUDGE RECIPE (10 sweets)

Sugar              220 g
Butter              30 g
Evaporated Milk    140 ml
```

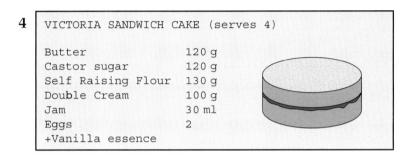

How much of each ingredient is needed to make:
a 20   **b** 15 sweets?

4

```
VICTORIA SANDWICH CAKE (serves 4)

Butter               120 g
Castor sugar         120 g
Self Raising Flour   130 g
Double Cream         100 g
Jam                   30 ml
Eggs                   2
+Vanilla essence
```

How much of each ingredient is needed to make a cake for:
a 2   **b** 6 people?

---

7 Scales and models

Example 1   A miniature railway has a model of a real train and carriages.
The real train and carriages are five times bigger than the models
(i.e. the scale is 1 to 5).

The model train is 3 m long. How long is the real train?

Length of real train = 5 × 3 m = 15 m

Exercise 7.1

1 a The model carriages are 4 m long. (Remember: the scale is 1 to 5.)
How long are the real carriages?

   b The chimney of the model train is 10 cm tall.
How tall is the chimney of the real train?

2 A child's toy TV is made using a scale of 1 to 6.

   a The toy TV is 8 cm wide.
How wide is the real TV?

   b The height of the toy TV is 6·5 cm.
How high is the real TV?

3 A model house is made on a scale of 1 to 20.
(The real house is 20 times bigger than the model.)

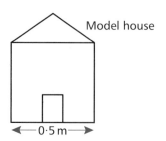
Model house

  a The door of the model is 3 cm wide.
    How wide is the door of the real house?

  b The model house is 0·5 m wide.
    How wide is the real house?

0·5 m

4 The plan of a playground is drawn using a scale of 1 cm to 2 m.

  a On the plan the playground is 30 cm long.
    How long is the actual playground?

  b On the plan the playground is 25 cm wide.
    How wide is the actual playground?

5 A map is drawn using a scale of 1 cm to 4 km.

  a What is the actual distance between Moortop and Lochhead?

  b On the map the loch is 0·5 cm long. How long is the actual loch?

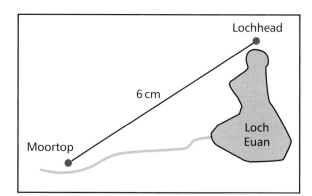

Lochhead

6 cm

Loch Euan

Moortop

Example 2 A model bicycle is 30 cm long.
              The real bicycle is 150 cm long.
              What scale was used to make the model bicycle?

              $150 \div 30 = 5$
              So the scale used is 1 cm to 5 cm or 1 to 5.

Exercise 7.2

1 An architect's model of a shopping centre is 2 m long.
The actual centre is to be 200 m long.
What is the scale of the model?   (Hint: $200 \div 2$)

2 A toy crane is 2 m tall.
It is a model of a real crane which is 18 m tall.
What scale was used to make the toy crane?

3 A child's toy piano stands 30 cm high.
It is a model of an actual piano which is 120 cm high.
What scale was used to make the toy piano?

4 On a map the distance between Glasgow and Edinburgh is 9 cm.
The actual distance between Glasgow and Edinburgh is 72 km.
How many kilometres does 1 cm on the map stand for?

5 A plan of a room is drawn.
On the plan the room is 40 cm long.
The actual room is 800 cm long.

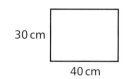

30 cm

40 cm

a What is the scale used on the plan?

b On the plan the room is 30 cm wide.
How wide is the actual room?

Investigation

Look at some boxes of model cars, boats and aircraft.
Find the scales the models are made to.
How tall would a model of you be if made to the same scale?

---

8 Graphs in proportion

Example 1 The graph shows the petrol used and the distance travelled by Phil
on his motorbike.

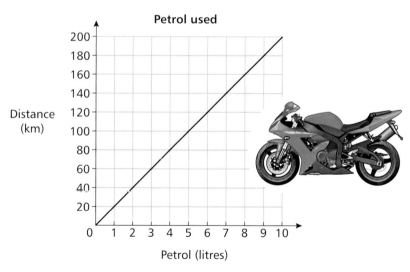

a How far will the bike go on 6 litres of petrol?
b How much petrol is needed to travel 60 km?

From the graph we see that:
a 120 km can be travelled on 6 litres
b 3 litres are needed for 60 km.

Exercise 8.1

1 This graph lets you change between miles and kilometres.

a Change 25 miles to kilometres.

b Estimate how many miles make 48 kilometres.

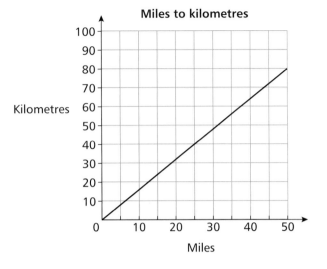

2 The graph shows one phone company's off-peak charges.

a What is the cost of a 40 minute call?

b How long do you get for 15p?

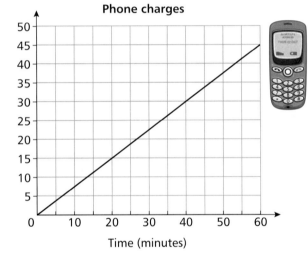

3 This graph lets you change between pounds and euro.

a Change £50 to euro.

b Change €105 to pounds.

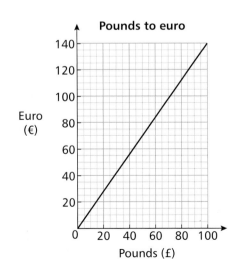

4 This graph converts between
   inches and centimetres.

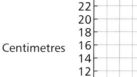

   a How many centimetres make
      i 6 inches
      ii 4 inches?
   b How many inches make
      i 20 cm
      ii 25 cm?

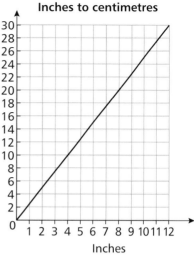

◄◄ RECAP

Rates

> **'per' means 'for each'**

A phrase that contains the word per is called a rate.

Examples of rates: cost per apple, miles per litre, kilometres per hour

Given, for example, the cost of one item, you can calculate the cost of several of that item.

Given the cost of more than one item you should be able to calculate the rate per item.

Best buy

You should be able to compare prices by calculating the cost per item (per unit) for each thing being compared.

Recipes

You should be able to increase and decrease ingredients in proportion to the number of portions required.

Scales

You should be able to use scales of models or maps to calculate lengths.

Proportion graphs

You should be able to read graphs, for example when changing from one currency to another.

1 Jodie jogs 200 metres in one minute.
How far will she jog in 4 minutes?

2 Roy reads 800 words in 10 minutes.
How many words does he read per minute?

3 Kelly has four films processed. It costs her £12.

  a What is the processing charge per film?

  b How much would it cost to have three films processed?

4 A tower of six toy bricks is 30 cm tall.
What height would a tower of ten of
the bricks reach?

5 6 organic eggs £1·20   10 organic eggs £1·80

  a Calculate the cost per egg for each box.

  b Which is the better buy?

6 When preparing to paint, Mac mixes 1 part thinners with 5 parts paint.

  a How much paint should he mix with 100 ml of thinners?

  b What quantity of thinners does he need with 10 litres of paint?

7 A 20 minute phone call costs 50p.
How much would a 24 minute call cost?

8

```
Gingerbread men recipe (20 men)

350 g      plain flour
2          tablespoons of ginger
100 g      margarine
180 g      brown sugar
4          tablespoons of syrup
```

How much of each ingredient is needed to make:

  a 10   **b** 30 gingerbread men?

9 Penny has a model of her horse made.
The model horse's back is 20 cm above the ground.
The actual height of the horse's back is 120 cm.

   a What is the scale of the model?

   b The tail on the model is 12 cm long.
   How long is the tail of the horse?

10 This graph lets you change between gallons and litres.

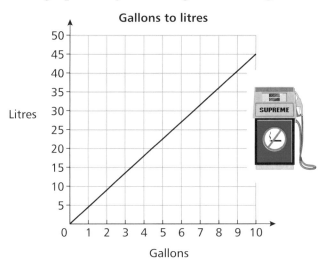

   a How many litres make 4 gallons?

   b Change 36 litres to gallons.

3 Perimeter and area

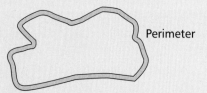

The perimeter of the racing circuit is the distance round it.

The area of the circuit is the surface it covers.

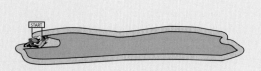

Perimeter

Area

The perimeter of the circuit is measured in metres (m).

The area of the circuit is measured in square metres (m²).

1 Review

◀◀ Exercise 1.1

1 The perimeter of the white sail is 60 + 80 + 100 cm = 240 cm. Calculate the perimeter of the blue sail.

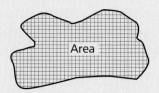

170   100
80 80
150   60

All measurements are in centimetres.

2 Calculate the perimeter of each sail.

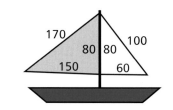

255   270
120   125
225   240

All measurements are in centimetres.

3 Which sports pitch has the greatest perimeter? The measurements are in metres.

a
85
65

Hockey pitch

b
82
112

Rugby pitch

c
103
84

Football pitch

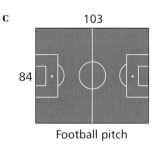

Memory jogger

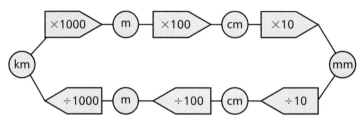

4 Change to metres:
  a 3 km   **b** 15 km

5 Express in metres:
  a 2000 mm   **b** 12 000 mm

6 Change to centimetres:
  a 140 mm   **b** 7 m

7 A length of copper wire is cut into three pieces.
These are 85 cm, 126 cm and 163 cm long.
What is the total length of wire in:
  a centimetres   **b** metres?

8 All the panes of glass in these two windows are the same size.

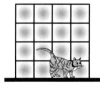 

  a Which window has the greater perimeter?
  How do you know?

  b Which window has the greater area?
  How do you know?

9 Find the area of each shaded shape by counting the squares.
Each square represents 1 square centimetre (1 cm²).
Write your answer like this: Area = ... cm².

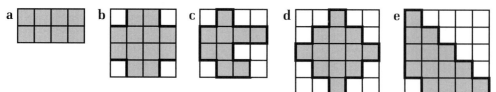

a   **b**   **c**   **d**   **e**

10 Estimate the area of each shape.
Only count a square if more than half of it is filled.
Each square represents an area of 1 cm².

A postage stamp   A calculator   A mobile phone

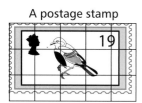
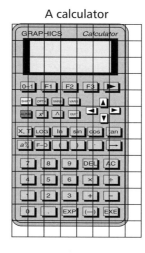

2 More on perimeter

Example   Calculate the perimeter of this shape.

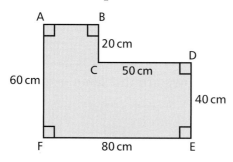

We first need to find the length of AB.
AB + CD = FE = 80 cm
CD = 50 cm so AB = 30 cm
Perimeter = 30 + 20 + 50 + 40 + 80 + 60
           = 280 cm

Exercise 2.1

1 Calculate the perimeter of each shape.

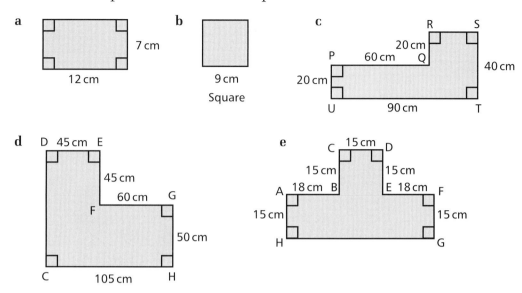

a 7 cm, 12 cm

b 9 cm, Square

c P Q R S T U, 20 cm, 60 cm, 20 cm, 40 cm, 90 cm

d D 45 cm E, 45 cm, 60 cm G, F, 50 cm, C 105 cm H

e C 15 cm D, 15 cm, 15 cm, A 18 cm B, E 18 cm F, 15 cm, 15 cm, H G

2 The diagram shows a plan of Dogood Academy, not drawn to scale.

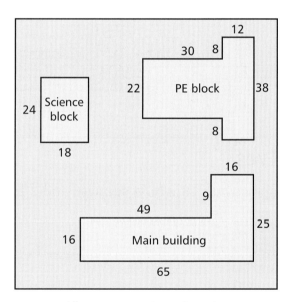

All measurements are in metres.

New guttering is being put right round the outside of each building.

What length of guttering is needed:
a round the Science block
b round the main building
c round the PE block
d altogether?

3 Measure the perimeter of each shape.

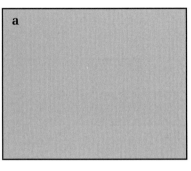

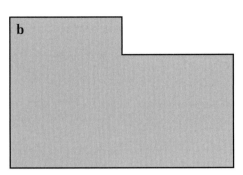

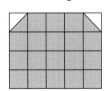

4 Which of these shapes has the greater perimeter?
Explain your answer.

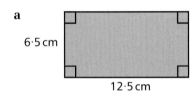

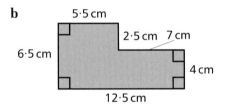

3 Area in halves

Example 1   This shape covers **18 whole squares** and **2 half squares**.

The 2 half squares make 1 whole square. + = ▨

Area of shape = 19 cm²

Example 2   The shaded shape has an area of 1 square because it covers half of a
two-square shape.

Exercise 3.1

1 Find the area of each of these shapes by counting squares and half squares.

Rectangle

Square

Pentagon

Octagon

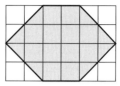
Hexagon

2 Now find the area of each shaded shape.

a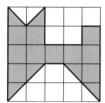

b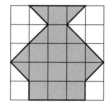

c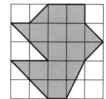

3 Find the area of these shapes. They are trickier!

a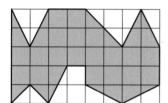

b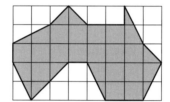

4 Area of irregular shapes

When a shape does not have straight edges, you can't find the *exact* area by counting squares.

> To get a good estimate of the area:
> - count the whole squares
> - count parts that are more than a half square as 1
> - count the half squares
> - don't count parts that are less than half a square.

This shape has an area of:
  7 full squares (*)
  4 more-than-half squares (•)
  2 half squares ($\frac{1}{2}$).
So the area is $7 + 4 + 1 = 12$ cm².
(Remember, this is not an exact answer.)

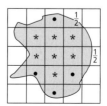
Each square represents 1 cm²

43

Exercise 4.1

In the maps of islands in questions **1–3**,
each square on the map stands for
1 square kilometre.

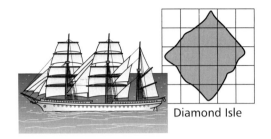

Diamond Isle

1 Check that Diamond Isle has an area of
   about 12 km².

2 Estimate the area of each of these islands in square kilometres (km²).

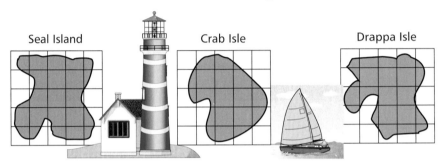

Seal Island   Crab Isle   Drappa Isle

3 a Estimate the area of Loch Cod.

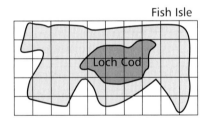

Fish Isle

Loch Cod

   b Estimate the area of land on Fish Isle.

4 You need a sheet of 1 cm squared paper.
   Let each square on the sheet represent an area of 1 km².
   a Draw an island which has an area of about 22 km².
   b Draw a loch on the island which has an area of about 7 km².

5 The city of Auchblane is shown on this map.
   The River Blane runs through the city.
   Each square on the grid represents an area of 1 km².

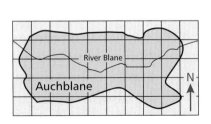

River Blane

Auchblane

N

   a Estimate the area of Auchblane.
   b Estimate the area of Auchblane to the north
      of the river
   c Estimate the area of the city to the south of the river.

5 The area of a rectangle

The area of this rectangle is 15 cm².

1	2	3	4	5
6	7	8	9	10
11	12	13	14	15

3 cm 5 cm

1	2	3	4	5
1	2	3	4	5
1	2	3	4	5

3 cm 5 cm

There are 5 squares in each row, and 3 rows of squares.
There are 5 × 3 squares altogether.

The area of the rectangle = 5 × 3 = 15 cm².

The area of any rectangle is found by multiplying its length and its breadth.

3 cm 5 cm

> **Area of rectangle = length × breadth**

Example 1 Calculate the area of this rectangle.

7 m
11 m

Area of rectangle
= length × breadth
= 11 m × 7 m
= 77 m².

Example 2 Find the area of this square.

8 cm
8 cm

Area of square
= length × breadth
= 8 cm × 8 cm
= 64 cm².
(Note that a square is a special kind of rectangle.)

Exercise 5.1

1 Calculate the area of each rectangle. All measurements are in centimetres.

a

2·4
2

b
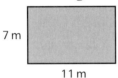
2·5
5

c

9·5
7

d

18
18

e
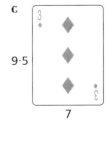
33
26

f
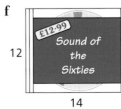
12
14

2 Donna has four posters on her wall.
Calculate the area of each poster. All measurements are in centimetres.

a

60

40

b

75

55

c

90

45

d

50

50

3 Lochside High School has four sizes of rectangular classrooms.
Sketches of them are shown below.
All measurements are in metres, so the areas will be in square metres (m²).
Arrange the classrooms in order of area, putting the largest one first.

a
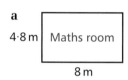
4·8 m | Maths room
8 m

b
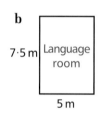
7·5 m | Language room
5 m

c
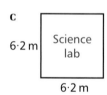
6·2 m | Science lab
6·2 m

d
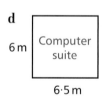
6 m | Computer suite
6·5 m

4 This is a plan of Anwar's garden.
  a Calculate the area of:
   i the grass
   ii the flower bed
   iii the path
  b i What is the length and breadth of the whole garden?
   ii What is the area of the whole garden?
  c i What is the length and breadth of the vegetable plot?
   ii Calculate the area of the vegetable plot.

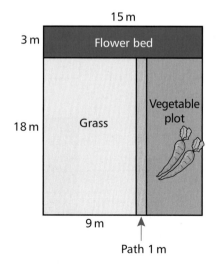

15 m

3 m | Flower bed

18 m | Grass | Vegetable plot

9 m

Path 1 m

Exercise 5.2

1 Measure the length and breadth of each rectangle.
Copy and complete the table.

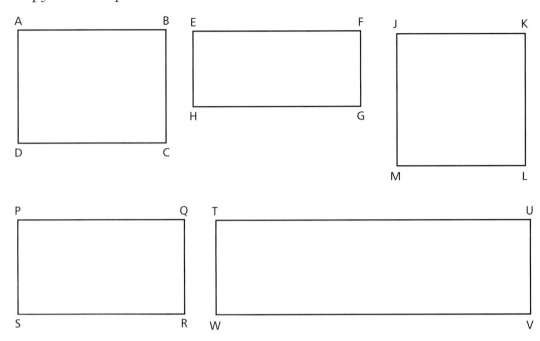

Rectangle	Length (cm)	Breadth (cm)	Area (cm²)
ABCD			
EFGH			
JKLM			
PQRS			
TUVW			

2 The diagram shows a plan of
Lochside High School, not drawn
to scale.

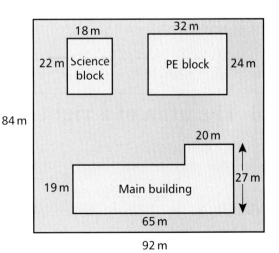

  a Calculate the area of:
  i   the Science block
  ii the PE block
  iii the main building.

  b What is the total area of the
  three buildings?

  c What is the area of the school
  grounds?

  d What area of the school grounds
  does not have a building on it?

3 The packet of Corny Pops is opened out flat to form the net of a cuboid.

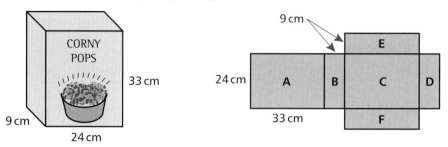

a Calculate the area of each of the faces A to F.

b What area of cardboard is needed altogether to make the packet of Corny Pops?

4 a Calculate the area of the small photograph.

b An enlargement of the photograph is made. The length and breadth have been doubled. Calculate the area of the enlargement.

c How many times greater is the area of the enlargement than the area of the original photograph?

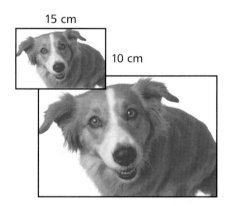

15 cm

10 cm

Challenge

This is a picture in its frame.
The frame is shaded.
The frame is 3 cm wide all the way round.
Calculate the area of the frame.

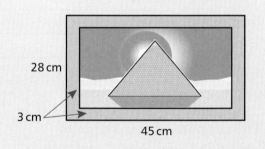

28 cm

3 cm

45 cm

6 The area of a right-angled triangle

Every right-angled triangle is half a rectangle.

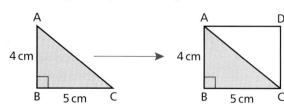

Triangle ABC is half of rectangle ABCD.

Area of rectangle ABCD = length × breadth
$$= 5 \times 4 = 20 \text{ cm}^2$$

Area of right-angled triangle ABC = $\frac{1}{2}$ of 20 cm²
$$= 10 \text{ cm}^2$$

> The area of a right-angled triangle is half the area of the surrounding rectangle.

Exercise 6.1

1 Write down the area of:
  a rectangle PQRS
  b triangle PQR.

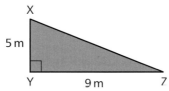

2 Write down the area of:
  a rectangle DEFG
  b triangle DEF.

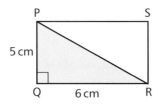

3 Calculate the area of ΔXYZ.

4 Calculate the area of each right-angled triangle.
   The measurements are all in centimetres.

a
8
14

b
7
11

c
15
20

d
9
7

e
14
24

f
27
7

g
3
5
4

h
13
5
12

5 Calculate the area of the blue part of the flag.

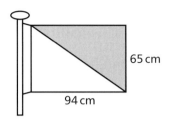

65 cm

94 cm

6 The diagram shows how three roads meet.
The angle between Ash Road and
Lang Drive is a right angle.
Calculate the area of the land surrounded
by the three roads.

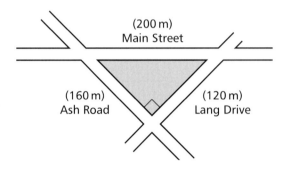

(200 m)
Main Street

(160 m)
Ash Road

(120 m)
Lang Drive

7 A farmer has 400 m of fencing.
He uses it to fence part of a field, as shown.
What area of field has he fenced off?

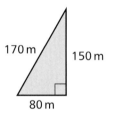

170 m

150 m

80 m

8 What area is shaded in each design?

a

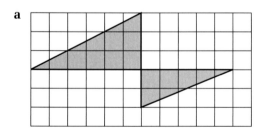

b

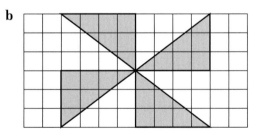

Each square represents 1 cm²

c

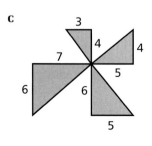

3

4

7

4

5

6

6

5

d

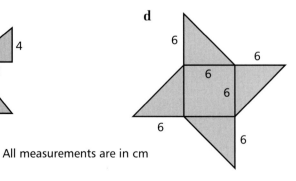

6

6

6

6

6

6

6

All measurements are in cm

Challenge

A kite is made up of four right-angled triangles.

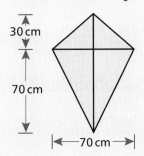

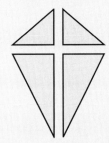

Find the area of this kite.

7 Problems with perimeter and area

It is important to be able to solve problems that involve perimeter and area.

Example   4 m² of carpet cost £36.
How much does it cost for:

a 1 m²   **b** 12 m²

of the carpet?

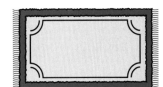

Area of carpet		Cost
4 m²	→	£36
a 1 m²	→	£36 ÷ 4 = £9
b 12 m²	→	£9 × 12 = £108

Exercise 7.1

1 8 m² of carpet cost £56.
How much does it cost for:

a 1 m²   **b** 15 m²?

2 An expensive carpet costs £26 per square metre.
How much does it cost for 18 m² of the carpet?

3 Moya runs six laps of the running track.
The distance she covers is 2400 metres.

a How far is one lap of the running track?

b Moya now runs eight laps of the track.
How far does she run this time?

4 Workmen can resurface 1600 m of a road in 5 days.
How many metres of road should they be able to resurface in:

a 1 day **b** 10 days?

5 A painter can paint 84 m² of a bridge in 7 hours.
What area of the bridge should he be able to paint in:

a 1 hour **b** 35 hours?

6 A farmer can plough 4500 m² of land in 8 hours.
Working at the same rate, how many square metres of land could he plough in 5 hours?

7 A racing driver can drive 9 laps of a circuit in 567 seconds.
How long would it take him to drive 48 laps of the circuit?

8 Three packets of grass seed are needed to cover an area of 72 m².
How many packets are needed to cover an area of 120 m²?

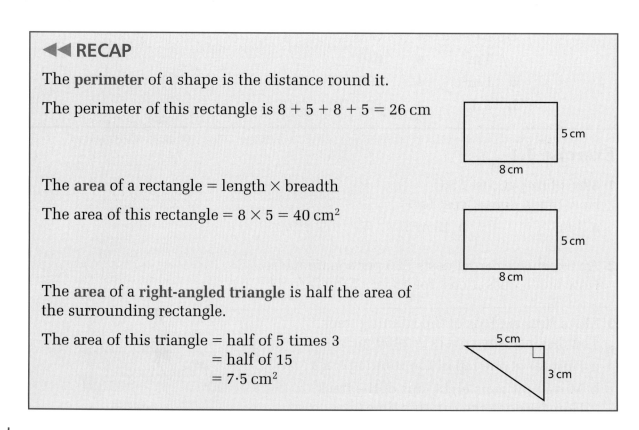

◀◀ **RECAP**

The **perimeter** of a shape is the distance round it.

The perimeter of this rectangle is 8 + 5 + 8 + 5 = 26 cm

The **area** of a rectangle = length × breadth

The area of this rectangle = 8 × 5 = 40 cm²

The **area** of a **right-angled triangle** is half the area of the surrounding rectangle.

The area of this triangle = half of 5 times 3
$$= \text{half of } 15$$
$$= 7{\cdot}5 \text{ cm}^2$$

1 Calculate the perimeter of:
 a the triangle
 b the rectangle.

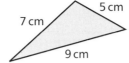

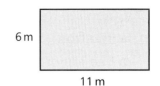

2 The diagram is part of a plan of a building.
 The measurements are in metres.
 a Calculate the length of AB.
 b Find the perimeter of the building.

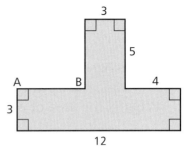

3 Find the area of this shape.
 Each square represents an area of 1 cm².

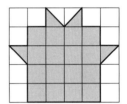

4 Each square of the grid represents an area of 1 m².
 Estimate the area of the pond by counting squares.

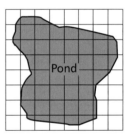

5 Calculate the area of each rectangle.

 a

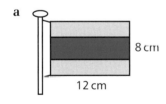

 b

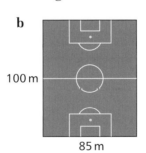

 c

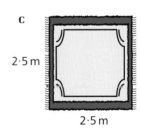

6 a Measure the length and
 breadth of this rectangle.
 b Calculate the area of
 the rectangle.

7 Calculate the area of:

  a the front of the packet

  b the top of the packet

  c the side of the packet.

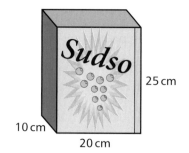

8 A full-sized snooker table is rectangular.
It is 12 feet long and 6 feet wide.
What is the area of a full-sized snooker table? (Take care with the units.)

9 The side of the ramp is a right-angled triangle.
Calculate the area of the side of the ramp.

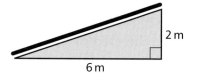

10 The two sails on the yacht are right-angled triangles.
Calculate the area of each sail.

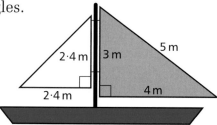

11 6 m² of carpet cost £120.
How much would 9 m² of the carpet cost?

4 Volume and weight

The **volume** of an object is how much space it takes up.

An apple usually has a greater volume than a strawberry.

Volume is measured in **cubic** units:

cubic millimetres (mm³)	cubic centimetres (cm³)	cubic metres (m³)

A grain of salt is about 1 cubic millimetre.

A dice is roughly 1 cubic centimetre.

Four wheelie-bins have a volume of roughly 1 cubic metre.

Weight is measured in **grams (g)**, **kilograms (kg)** and **tonnes**.

A cubic centimetre of pure water weighs exactly 1 gram.

A litre of pure water weighs exactly 1 kilogram.

A cubic metre of pure water weighs exactly 1 tonne.

A drawing pin weighs about a gram.

A bag of sugar weighs 1 kilogram.

Tyle and Lale White Sugar 1 kg

A car weighs about a tonne.

1 Review

◀◀ Exercise 1.1

1 This shape is made up of nine cubes.
Each cube represents 1 cubic centimetre.
The volume of the shape is 9 cm³.

Count the cubes in each shape below to find the volume of the shape.
Write your answer like this: volume of **a** = ... cm³.

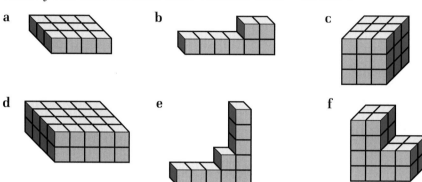

a   **b**   **c**

d   **e**   **f**

2 Which of the solids in question **1** takes up:
   a the smallest space   **b** the most space?

3 Which unit of **volume** would you use to measure:
   a a packet of breakfast cereal
   b a diamond
   c a warehouse
   d a raindrop
   e the water in a swimming pool?

4 Which unit of weight would you use to measure these objects?

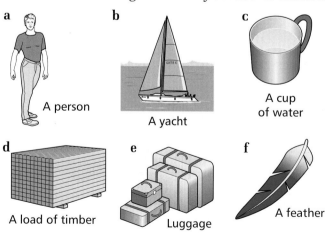

a A person   **b** A yacht   **c** A cup of water

d A load of timber   **e** Luggage   **f** A feather

5 Which object in question **4** is:
   a the heaviest   **b** the lightest?

2 Finding the volume of a cuboid by counting cubes

This solid is made up of two identical layers of centimetre cubes.

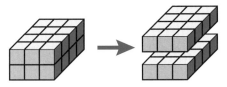

The total number of cubes is found by multiplying the number of cubes in a layer by the number of layers.

The number of cubes in a layer = 12   (Each layer has 4 rows of 3 cubes = 12 cubes)
The number of layers = 2
So the number of cubes = 12 × 2 = 24
Volume of solid = 24 cm³

Exercise 2.1

1 a Write down how many cubes are in each layer.
 b How many layers are there?
 c Multiply to find the volume of the cuboid in cubic centimetres (cm³).

2 Follow the instructions in question **1** to find the volumes of these cuboids.

a 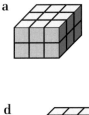   b 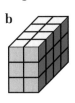   c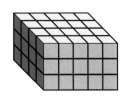

d 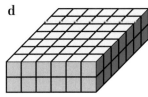   e 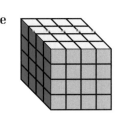   f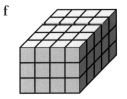

g 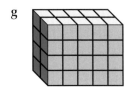   h 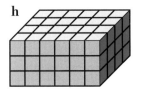   i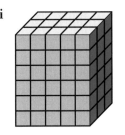

3 a i Check that the number of cubes in each layer is 4×3.
　ii How many layers are there?
　iii What is the volume of the cuboid in cm³?
b i What is the length of the cuboid?
　ii What is its breadth?
　iii What is its height?
　iv Calculate length $\times$ breadth $\times$ height.
c What do you notice about your answers to **a iii** and **b iv**?

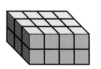

---

3 Finding the volume of a cuboid using a formula

This cuboid is 4 cm by 3 cm by 2 cm.
To find its volume without counting cubes, we use the
formula:

volume of cuboid = length $\times$ breadth $\times$ height

Volume = 4 cm $\times$ 3 cm $\times$ 2 cm = 24 cm³.

Example　Use the formula to calculate the volume
of this cuboid.
Volume $= l \times b \times h$
$= 6 \times 3 \times 4$
$= 72$ cm³

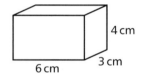

4 cm
3 cm
6 cm

Exercise 3.1

1 A packet of rice flakes is 25 cm by 10 cm by 35 cm.
Use the formula to calculate the volume of the packet.

RICE FLAKES

35 cm
10 cm
25 cm

2 Find the volume of each of these.
All measurements are in centimetres.

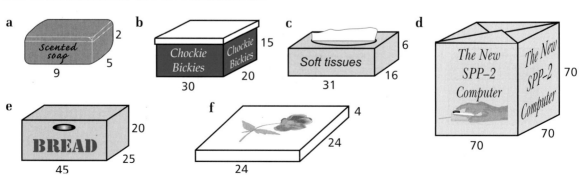

a Scented soap — 2, 5, 9
b Chockie Bickies — 15, 20, 30
c Soft tissues — 6, 16, 31
d The New SPP-2 Computer — 70, 70, 70
e BREAD — 20, 25, 45
f 4, 24, 24

3 A shoe box is 30 cm by 19 cm by 9 cm.
Calculate the volume of the shoe box.

> Large objects with length, breadth and height given in metres have their volume measured in **cubic metres** (m^3).

4 Calculate the volume of the garage.

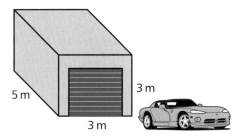

5 m   3 m   3 m

5 The diagrams show the sizes of some rooms in a school.
Work out the volume of each room in cubic metres.

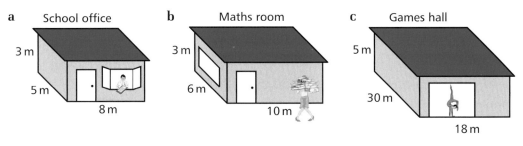

a School office 3 m 5 m 8 m

b Maths room 3 m 6 m 10 m

c Games hall 5 m 30 m 18 m

6 A rabbit hutch is in the shape of a cuboid.
It is 4 metres by 2 metres by 1·5 metres.
Calculate the volume of the rabbit hutch.

> Small objects with length, breadth and height given in millimetres have their volume measured in **cubic millimetres** (mm^3).

7 Calculate the volume of the pencil sharpener.

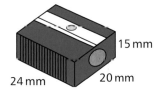

15 mm   20 mm   24 mm

8 Calculate the volume of each of these objects in mm^3.

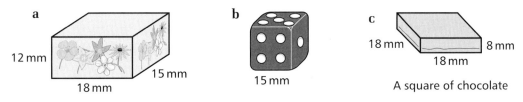

a 12 mm 15 mm 18 mm

b 15 mm

c 18 mm 8 mm 18 mm
A square of chocolate

9 A contact lens case is in the shape of a cuboid.
It is 22 mm by 24 mm by 9 mm.
Calculate the volume of the case.

Exercise 3.2

1 What volume of water will the basin hold?

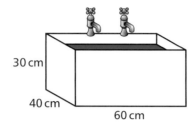

2 Sharon has an earring made from three silver cubes.

   a Calculate the volume of each cube in the earring.

   b What volume of silver is needed to make one *pair* of the earrings?

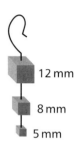

3 A video case is 20 cm by 10 cm by 2·5 cm.

   a What is the volume of the video case?

   b Twelve of the cases are put side by side on a shelf. What volume of space do the twelve cases occupy?

4 The dimensions of two fish tanks are shown.

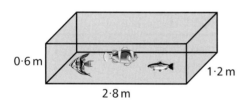 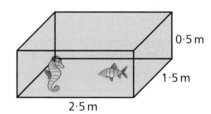

Which tank has the greater volume? By how much?
Show all your working.

5 The table suggests some cuboids that can be measured in your classroom. Copy the table, measure the items and calculate their volumes. Add some other cuboids of your own to the table.

Cuboid	Length	Breadth	Height	Volume
Maths book	... cm	... cm	... cm	... cm^3
Classroom	... m	... m	... m	... m^3
Filing cabinet	... cm	... cm	... cm	... cm^3
Metal cupboard	... cm	... cm	... cm	... cm^3

4 The volume of liquids

The volume of a liquid is usually measured in **litres**.

| 1000 cubic centimetres = 1 litre (ℓ) |

A cubic centimetre is usually called a **millilitre** (ml).
On this jug, each litre is divided into 5 small divisions.

1 litre ÷ 5 = 1000 ml ÷ 5 = 200 ml

Each small division is 200 ml.
This jug contains 800 ml of liquid.

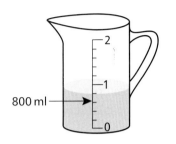

Exercise 4.1

1 Write down the volume of liquid, in millilitres, in each of these jugs.

a  **b**  **c**

2 Write down the volume of liquid, in litres and millilitres, in each of these jugs.

a  **b**  **c**

3 a On this jug, each litre is divided into how many small divisions?

  b How many millilitres are there in each small division?

  c What volume of liquid is in the jug, in millilitres?

4 Write down the volume of liquid in each jug, in litres and millilitres.

a  **b**  **c**

5 A carton contains 750 ml of lime juice.
Chloe drinks 180 ml of the lime juice.
How much is left?

6 A full bottle holds 1 litre of cola.
  a How many millilitres is this?
  b Jeff drinks 230 ml of the cola.
     How much is left?

7 Wasif makes 5 litres of peach punch for a party.
Each of his glasses can hold 200 ml of punch.
How many full glasses of peach punch can he pour?

8 A bottle contains 250 ml of water.
A case holds 24 bottles of the water.
  a How much water is in a case?
  b Write this volume in litres.

9 The instructions on Bill's bottle of cough mixture read:
'Two 5 ml spoonfuls three times a day'.
  a How many spoonfuls of medicine should Bill take each day?
  b How many millilitres of medicine should Bill take each day?
  c The bottle holds 450 ml of cough mixture.
     For how many days will Bill's medicine last?

Cough Mixture

Exercise 4.2   Reminder: 1000 millilitres = 1 litre

1 Change these volumes to litres.
  a 3000 ml   **b** 7000 ml   **c** 16 000 ml   **d** 25 000 ml

2 Change these volumes to millilitres.
  a 4 litres   **b** 9 litres   **c** 13 litres   **d** 21 litres

3 Change these volumes to litres and millilitres.
  a 1500 ml   **b** 2100 ml   **c** 6750 ml   **d** 17 250 ml

4 Change these volumes to millilitres.
  a 1 litre 700 ml   **b** 3 litres 650 ml   **c** 14 litres 940 ml

5 Change these to litres and millilitres.
  a 2·5 litres   **b** 3·4 litres   **c** 8·7 litres   **d** 4·25 litres

6 Change to litres.
  a 3 litres 500 ml   **b** 5 litres 100 ml   **c** 9 litres 750 ml

Challenge

A supermarket sells ice cream in different sizes of carton, as shown below.

Kool
Ice Cream
2 litres

F
A
B
300 ml

ACE
CREAM
850 ml

NICE ICE NICE ICE
1 litre

Hot Ice
1·5 litres

1 Put the cartons in order, with greatest volume first.
2 How many millilitres are there in the Hot Ice carton?
3 What is the total volume of ice cream in the five cartons?
  Give your answer in litres and millilitres.
4 We can write 300 ml of Fab ice cream as 0·3 litre.
  Write 850 ml of Ace Cream in the same way.

5 Weight: reading scales

> 1 kilogram = 1000 grams

These scales measure kilograms (kg) and grams (g).

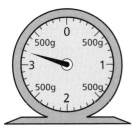

Each kilogram is divided
into 10 small parts.

Each kilogram is divided
into 4 small parts.

Each kilogram is divided into
10 small parts.

$1000 \div 10 = 100 \, g$
Each small part is 100 g.

The arrow is pointing to 3 kg 200 g.

Each kilogram is divided
into 4 small parts.

$1000 \div 4 = 250 \, g$
Each small part is 250 g.

The arrow is pointing to 1 kg 750 g.

Remember to check what each small part is worth.

Exercise 5.1

1 Write down the weights, in kilograms and grams, shown on these scales.

Be careful!

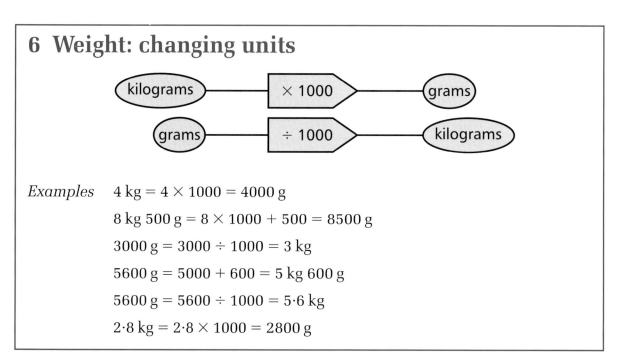

6 Weight: changing units

kilograms —— × 1000 —— grams

grams —— ÷ 1000 —— kilograms

Examples   4 kg = 4 × 1000 = 4000 g

8 kg 500 g = 8 × 1000 + 500 = 8500 g

3000 g = 3000 ÷ 1000 = 3 kg

5600 g = 5000 + 600 = 5 kg 600 g

5600 g = 5600 ÷ 1000 = 5·6 kg

2·8 kg = 2·8 × 1000 = 2800 g

Exercise 6.1

1 Change these weights into grams.

    a 3 kg   **b** 7 kg   **c** 12 kg   **d** 19 kg

    e 4 kg 500 g   **f** 1 kg 250 g   **g** 7 kg 340 g   **h** 5 kg 900 g

    i 2 kg 100 g   **j** 1 kg 90 g   **k** 8 kg 50 g   **l** 3 kg 75 g

2 Change these weights to kilograms and grams.

    a 5000 g   **b** 9000 g   **c** 6500 g

    d 3700 g   **e** 12 425 g   **f** 5075 g

    g 4050 g   **h** 8205 g

3 Change these weights into kilograms.

    For example: 6350 g = 6350 ÷ 1000 = 6·35 kg

    a 2100 g   **b** 5400 g   **c** 18 300 g   **d** 14 250 g

    e 1350 g   **f** 3625 g   **g** 500 g   **h** 985 g

4 Change these weights into grams (by multiplying by 1000).

    a 1·5 kg   **b** 3·7 kg   **c** 6·25 kg   **d** 8·125 kg

    e 15·7 kg   **f** 18·25 kg   **g** 0·6 kg   **h** 0·75 kg

5 Change these weights into grams.

    For example: 3 kg 450 g = 3000 + 450 = 3450 g

    a 6 kg 500 g   **b** 2 kg 200 g   **c** 7 kg 450 g

    d 17 kg 300 g   **e** 12 kg 840 g   **f** 5 kg 375 g

    g 20 kg 150 g   **h** 5 kg 25 g   **i** 1 kg 50 g

7 Using weight

It is important to be able to solve problems that involve weight.

Remember:
> 1000 g = 1 kg
> 1000 kg = 1 tonne

Exercise 7.1

1 State which unit you would use to weigh:

    a yourself   **b** a bar of chocolate   **c** a heavy suitcase

    d a bus   **e** a slice of bread   **f** the water in a swimming pool

2 Here is what is in Mrs Boyd's shopping trolley:

450 g 3 kg 150 g 680 g

570 g 1·6 kg 800 g

a How many grams of flour did she buy?

b How much heavier is the bread than the jam?

c What is the lightest item she bought?

d What is the total weight of Mrs Boyd's shopping?

3 Sam is making corned beef hash for three people.
The main ingredients are:

    60 g butter

    750 g potatoes

    325 g corned beef

What is the total weight of the main ingredients:

a in grams

b in kilograms and grams

c in kilograms?

4 Four friends had their luggage weighed at an airport.

    Lucy: $16\frac{1}{2}$ kg

    Craig: 14 kg 750 g

    Leah: 13·4 kg

    Joe: 17 kg 100 g

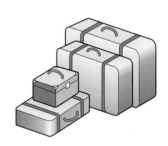

a Whose luggage was heaviest?

b How much heavier was Craig's luggage than Leah's?

c What was the total weight of the luggage?
Give your answer in kilograms and grams.

Exercise 7.2

1 Three packets of crisps weigh 75 grams.
What is the weight of:

   a 1 packet of the crisps

   b 5 packets of the crisps?

2 To make 16 fruit tartlets, 240 grams of plain flour are needed.
How many grams of flour are needed to make:

   a 1 tartlet   **b** 10 tartlets   **c** 24 tartlets?

3 Four bottles of water weigh 6 kilograms.

   a What is the weight of one bottle of water?

   b What is the weight of three of the bottles of water?

4 Two packs of candles weigh 800 grams.
What is the weight of three packs of the candles?

5 Six cans of tomato soup weigh 2400 grams.
What is the weight of nine of the cans?

6 Two bananas are needed to make banoffi pie for six people.
How many bananas are need to make the pie for 15 people?

7 Two large bags of flour weigh 9 kilograms.
What is the weight of three bags of the flour?

8 Ten copies of the *Daily Echo* newspaper weigh 840 grams.
What is the weight of 100 copies of the newspaper?
Give your answer in kilograms and grams.

Challenge

1 Which of these is heavier?

   a A sackful of potatoes or a sackful of mushrooms?

   b A tonne of feathers or a tonne of bricks?

2 Which has the greater volume?

   a A kilogram of golf balls or a kilogram of table tennis balls?

   b A litre of water or a litre of air?

◀◀ RECAP

The **volume** of a cuboid = length × breadth × height

$$V = l \times b \times h$$

The volume of this cuboid $= l \times b \times h$
$$= 8 \times 5 \times 4$$
$$= 160 \text{ cm}^3$$

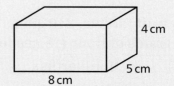

The volume of a liquid is usually measured in **litres** (ℓ) and **millilitres** (ml).

1000 cubic centimetres = 1 litre

A cubic centimetre is often called a millilitre.

Sugar lump
1 ml

1 litre

1000 ml = 1 litre

The main units of weight are the gram (g), kilogram (kg) and tonne.

1000 g = 1 kg
1000 kg = 1 tonne

1 gram

Tyle
and
Lale
White Sugar
1 kg

1 tonne

1 kilogram

You should be able to change between units of weight:

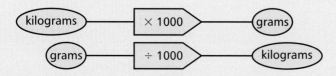

kilograms — × 1000 — grams

grams — ÷ 1000 — kilograms

Examples   5 kg = 5 × 1000 = 5000 g
3 kg 600 g = 3 × 1000 + 600 = 3600 g
7000 g = 7000 ÷ 1000 = 7 kg
2800 g = 2000 + 800 = 2 kg 800 g
2800 g = 2800 ÷ 1000 = 2·8 kg
3·5 kg = 3·5 × 1000 = 3500 g

1 What is the volume of this shape?
Each small cube has a volume of 1 cm³.

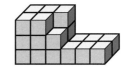

2 What is the volume of this cuboid?
Each small cube has a volume of 1 cm³.

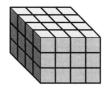

3 **a** Calculate the volume of the microwave oven.
b Calculate the volume of the box that the
microwave oven was delivered in.

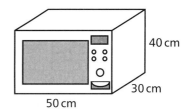

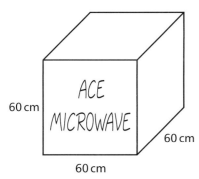

c How much greater is the volume of the box than
the volume of the microwave oven?

4 What volume of liquid is in the jug, in litres
and millilitres?

5 A bottle holds 2·6 litres of orange juice.
How many cups of orange juice can be filled
from the bottle if each cup holds 200 millilitres?

6 What weight, in kilograms and grams,
is shown on the scale?

7 Change these weights to kilograms.
   a 6000 grams   **b** 17 000 grams   **c** 4500 grams

8 Change these weights to grams.
   a 5 kilograms   **b** 3·2 kilograms   **c** 8·65 kilograms

REVISE

9 Change these weights to kilograms and grams.

   a 2·6 kilograms   **b** 9·25 kilograms   **c** 6350 grams

10 These scales are balanced with eight
   small parcels on one side and a weight
   of 2·8 kg on the other side.
   Each small parcel weighs the same.
   Calculate the weight of one small parcel.

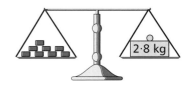

11 These three full cans contain 4·5 litres
   of ginger beer altogether.
   How much ginger beer will five full
   cans contain?

5 Fractions and percentages

The Romans did not use numerals to represent fractions. Words were used to indicate parts of a whole.

The Egyptians used fractions but only one of any type: one half, one third, one quarter …
but not three-quarters.

This chapter looks at such fractions.

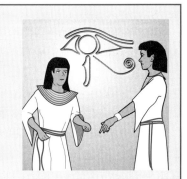

1 Review

◄◄ Exercise 1.1

1 Calculate:

   a $2347 + 324$ **b** $5296 + 1687$ **c** $3493 - 268$ **d** $4823 - 2962$

2 Multiply:

   a 32×2 **b** 46×3 **c** 127×5 **d** 437×4

   e 3261×5 **f** 3947×2 **g** 2849×3 **h** 5948×4

   i 269×6 **j** 841×7 **k** 81×9 **l** 397×8

   m 1245×6 **n** 2837×7 **o** 3821×9 **p** 8741×8

3 Work out:

   a $239 \div 2$ **b** $843 \div 3$ **c** $295 \div 5$ **d** $896 \div 4$

   e $3846 \div 3$ **f** $6685 \div 5$ **g** $6532 \div 4$ **h** $8576 \div 2$

   i $455 \div 7$ **j** $600 \div 8$ **k** $564 \div 6$ **l** $783 \div 9$

   m $2550 \div 6$ **n** $3896 \div 8$ **o** $6752 \div 8$ **p** $2511 \div 9$

4 What is:

   a $89 \div 2$ **b** $233 \div 4$ **c** $1241 \div 5$ **d** $1971 \div 4$

   e $8 \cdot 8 \div 2$ **f** $6 \cdot 9 \div 3$ **g** $1 \cdot 6 \div 4$ **h** $8 \cdot 24 \div 4$

   i $1 \cdot 35 \div 5$ **j** $2 \cdot 34 \div 6$ **k** $31 \cdot 5 \div 7$ **l** $34 \cdot 8 \div 8$?

5 Calculate:

    a 23×10 **b** 794×100 **c** 296×10 **d** $29{\cdot}7 \times 10$

    e $1{\cdot}23 \times 100$ **f** 867×100 **g** $8{\cdot}7 \times 100$ **h** $0{\cdot}63 \times 10$

    i $327 \div 10$ **j** $690 \div 10$ **k** $72 \div 10$ **l** $820 \div 100$

    m $9{\cdot}7 \div 10$ **n** $8 \div 100$ **o** $62 \div 100$ **p** $0{\cdot}27 \div 100$

6 Tom bought a pair of training shoes for £79·99.
In another shop, he noticed the same pair of trainers for £4·50 less.
How much did the trainers cost in the second shop?

7 Jane walked for 3 hours 25 minutes to reach the top of a hill.
It took her 40 minutes less to walk back down the hill.
How much time did she spend walking altogether?

8 Kim's gran gave Kim £6·50 each month for pocket money.
Her parents gave her four times as much.
How much was she given in total each month?

2 Egyptian fractions

Egyptian fractions have a **numerator** (top number) of 1, for example $\frac{1}{2}$, $\frac{1}{10}$.
Here are two examples of calculations with these fractions.

Example 1   Calculate $\frac{1}{3} \times 48$.
           $\frac{1}{3} \times 48 = 48 \div 3 = 16$

Example 2   Find $\frac{1}{4}$ of 60.
           $\frac{1}{4}$ of $60 = 60 \div 4 = 15$ 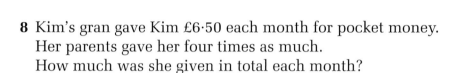

'of' means '$\times$'

Exercise 2.1

1 Find:

    a $\frac{1}{2}$ of 50   **b** $\frac{1}{3}$ of 72   **c** $\frac{1}{4}$ of 80   **d** $\frac{1}{3}$ of 66

    e $\frac{1}{4}$ of 100   **f** $\frac{1}{2}$ of 86   **g** $\frac{1}{3}$ of 84   **h** $\frac{1}{4}$ of 64

    i $\frac{1}{2}$ of 150   **j** $\frac{1}{4}$ of 160   **k** $\frac{1}{3}$ of 240   **l** $\frac{1}{2}$ of 424

    m $\frac{1}{2}$ of 830   **n** $\frac{1}{4}$ of 900   **o** $\frac{1}{3}$ of 864   **p** $\frac{1}{2}$ of 584

    q $\frac{1}{3}$ of 783   **r** $\frac{1}{4}$ of 544   **s** $\frac{1}{3}$ of 777   **t** $\frac{1}{4}$ of 676

2 Calculate:

   a $\frac{1}{2} \times 4980$ **b** $\frac{1}{3} \times 3780$ **c** $\frac{1}{4} \times 8920$ **d** $\frac{1}{2} \times 12464$

   e $\frac{1}{3} \times 3243$ **f** $\frac{1}{4} \times 8328$ **g** $\frac{1}{2} \times 6980$ **h** $\frac{1}{3} \times 6786$

   i $\frac{1}{2} \times 6840$ **j** $\frac{1}{3} \times 4380$ **k** $\frac{1}{4} \times 4064$ **l** $\frac{1}{2} \times 8642$

3 Work out:

   a $\frac{1}{2} \times 1\cdot2$ **b** $\frac{1}{3} \times 1\cdot5$ **c** $\frac{1}{4} \times 2\cdot4$ **d** $\frac{1}{2} \times 3\cdot6$

   e $\frac{1}{3} \times 3\cdot24$ **f** $\frac{1}{4} \times 8\cdot36$ **g** $\frac{1}{2} \times 9\cdot16$ **h** $\frac{1}{3} \times 5\cdot43$

   i $\frac{1}{2} \times 10\cdot38$ **j** $\frac{1}{4} \times 12\cdot48$ **k** $\frac{1}{3} \times 13\cdot53$ **l** $\frac{1}{4} \times 15\cdot64$

4 Find:

   a $\frac{1}{8}$ of 168   **b** $\frac{1}{9}$ of 288   **c** $\frac{1}{7}$ of 98   **d** $\frac{1}{6}$ of 96

   e $\frac{1}{7}$ of 315   **f** $\frac{1}{8}$ of 592   **g** $\frac{1}{7}$ of 441   **h** $\frac{1}{9}$ of 675

   i $\frac{1}{9}$ of 882   **j** $\frac{1}{8}$ of 504   **k** $\frac{1}{7}$ of 2842   **l** $\frac{1}{6}$ of 5376

   m $\frac{1}{8}$ of 3664   **n** $\frac{1}{9}$ of 5913   **o** $\frac{1}{6}$ of 5112   **p** $\frac{1}{9}$ of 6777

5 Calculate:

   a $\frac{1}{6}$ of 19·2   **b** $\frac{1}{8}$ of 27·6   **c** $\frac{1}{9}$ of 57·06   **d** $\frac{1}{7}$ of 45·15

   e $\frac{1}{7}$ of 16·45   **f** $\frac{1}{8}$ of 52·32   **g** $\frac{1}{6}$ of 28·14   **h** $\frac{1}{9}$ of 78·03

6 It costs £7·50 for an adult to watch the local rugby team play.
   A child ticket costs one third of this.
   How much does it cost for a child's ticket?

7 There are 36 desks in the maths classroom.
   The study room has one quarter this number of desks.
   How many desks are in the study room?

8 Paul drove 740 miles in one week.
   The next week, he drove one fifth of this distance.
   How many miles did he drive in the second week?

9 In a sale, the price of a coat was reduced by one third.
   a By how much was the price reduced?
   b What was the new price of the coat?

£127·80

Reduced by
1/3

10 In a long distance race at school, the best senior pupil time was 28 minutes.
    The fastest junior pupil time was one quarter slower than this.
    What was the junior pupil time?

Using a calculator to find fractions

Example 1   Calculate $\frac{1}{12}$ of 324.

Enter: $\boxed{324}\ \boxed{\div}\ \boxed{12}\ \boxed{=}$

The calculator reads $\boxed{\qquad 27 \qquad}$

So $\frac{1}{12}$ of 324 = 27

Example 2   Find $\frac{1}{15}$ of £35·34. Round your answer to the nearest penny.

Using a calculator, $\frac{1}{15}$ of £35·34 = 35·34 ÷ 15 = £2·356

So, to the nearest penny, $\frac{1}{15}$ of £35·34 = £2·36

Exercise 2.2

1 Use your calculator to find:

  a $\frac{1}{11}$ of 462   **b** $\frac{1}{12}$ of 636   **c** $\frac{1}{12}$ of 588   **d** $\frac{1}{13}$ of 676

  e $\frac{1}{14}$ of 504   **f** $\frac{1}{11}$ of 957   **g** $\frac{1}{15}$ of 630   **h** $\frac{1}{14}$ of 742

  i $\frac{1}{20}$ of 1060   **j** $\frac{1}{25}$ of 1025   **k** $\frac{1}{50}$ of 2600   **l** $\frac{1}{24}$ of 1296

  m $\frac{1}{40}$ of 2240   **n** $\frac{1}{20}$ of 3850   **o** $\frac{1}{60}$ of 1440   **p** $\frac{1}{80}$ of 5120

2 Now do these on your calculator:

  a $\frac{1}{11} \times 49·5$ **b** $\frac{1}{13} \times 72·8$ **c** $\frac{1}{14} \times 85·4$

  d $\frac{1}{20} \times 68$ **e** $\frac{1}{18} \times 120·6$ **f** $\frac{1}{25} \times 115$

  g $\frac{1}{18} \times 100·8$ **h** $\frac{1}{19} \times 142·5$ **i** $\frac{1}{24} \times 211·2$

  j $\frac{1}{35} \times 234·5$ **k** $\frac{1}{17} \times 168·3$ **l** $\frac{1}{23} \times 170·2$

> '×' can be swapped for 'of'

3 Find:

  a $\frac{1}{15}$ of £48·60   **b** $\frac{1}{12}$ of £76·20   **c** $\frac{1}{18}$ of £98·28   **d** $\frac{1}{24}$ of £82·80

  e $\frac{1}{26}$ of £91   **f** $\frac{1}{13}$ of £44·98   **g** $\frac{1}{16}$ of 106·88   **h** $\frac{1}{25}$ of 97·50

4 Mr and Mrs Sahwar had a total of 583 miles to drive to their destination.
After one hour, they had driven $\frac{1}{11}$ of the way.
How far had they driven after one hour?

5 A builder had 2880 bricks delivered to build an extension.
Unfortunately, $\frac{1}{12}$ of them were damaged.
How many bricks were damaged?

6 Kirsty's last maths exam was made up of two papers.
The total number of possible marks was 144.
She lost $\frac{1}{6}$ of the total marks in the first paper.

a How many marks did she lose in the first paper?

b She lost $\frac{1}{16}$ of the total marks in the second paper.

How many marks did she lose in the second paper?

c What was her final total mark?

7 A local council inspected 2550 houses to see if their loft insulation was
satisfactory.
$\frac{1}{15}$ of these houses had unsatisfactory insulation.

a How many houses had unsatisfactory insulation?

b How many had satisfactory insulation?

8 The Eiffel Tower is 300 metres high.
A souvenir model is $\frac{1}{1600}$ of the actual size.
How tall is the souvenir model?

3 Percentages

You will find many examples of percentages in everyday life.
Percent means out of a hundred.
The sign % is often used for per cent.

So 3% means 3 out of 100. This is written as $\frac{3}{100}$.

8% means the same as $\frac{8}{100}$.

Percentages can be written as decimals or fractions.

$1\% - \frac{1}{100} - 1 \div 100 - 0·01$

> To find 1%, divide by 100.

Example 1   Find 1% of £48.

$48 \div 100 = 0·48$

So 1% of £48 = £0·48

Example 2   Find 3% of £60.

1% of 60 = 0·60

So 3% of £60 = 0·60 × 3 = £1·80

Some percentages are easy to find if you treat them as common fractions.

50 is half of 100	... $50\% = \frac{1}{2}$
$33\frac{1}{3}$ is a third of 100	... $33\frac{1}{3}\% = \frac{1}{3}$
25 is a quarter of 100	... $25\% = \frac{1}{4}$
20 is a fifth of 100	... $20\% = \frac{1}{5}$
10 is a tenth of 100	... $10\% = \frac{1}{10}$
1 is a hundredth of 100	... $1\% = \frac{1}{100}$

Learn these ones.

Example 3 Find 10% of £90.

10% of $90 = \frac{1}{10}$ of $90 = 90 \div 10 = 9$

So 10% of £90 = £9

Example 4 Find $33\frac{1}{3}\%$ of £90.

$33\frac{1}{3}\%$ of $90 = \frac{1}{3}$ of $90 = 90 \div 3 = 30$

So $33\frac{1}{3}\%$ of £90 = £30

Example 5 Find 25% of £84.

25% of $84 = \frac{1}{4}$ of $84 = 84 \div 4 = 21$

So 25% of £84 = £21

Exercise 3.1

1 Write each percentage as a fraction out of 100.

   a 25%   **b** 60%   **c** 80%   **d** 90%   **e** 99%   **f** 7%

2 Write each fraction as a percentage.

   a $\frac{30}{100}$ **b** $\frac{70}{100}$ **c** $\frac{85}{100}$ **d** $\frac{93}{100}$ **e** $\frac{6}{100}$ **f** $\frac{11}{100}$

3 Write each percentage as a fraction in its simplest form.

   a 25%   **b** 10%   **c** 20%   **d** $33\frac{1}{3}\%$ **e** 50%   **f** 1%

4 Write each percentage as a decimal, e.g. $20\% = 20 \div 100 = 0{\cdot}2$.

   a 25%   **b** 40%   **c** 60%   **d** 80%

5 Write each decimal as a percentage, e.g. $0{\cdot}2 = 0{\cdot}2 \times 100\% = 20\%$.

   a 0·3   **b** 0·7   **c** 0·9

6 Find 1% of:

 a 50   **b** 87   **c** 92   **d** 6   **e** 53

 f 234   **g** 5·4   **h** 604   **i** 670   **j** 7·4

7 Calculate these by first finding 1%:

 a 2% of 30   **b** 4% of 80   **c** 3% of 52   **d** 5 % of 150

 e 3% of 48   **f** 5% of 160   **g** 4% of 250   **h** 6% of 96

8 Find 10% of:

 a 49   **b** 38   **c** 94   **d** 230

9 Calculate:

 a 50% of 40   **b** 50% of 80   **c** 50% of 260   **d** 50% of 400

 e 50% of 600   **f** 50% of 380   **g** 50% of 440   **h** 50% of 660

10 Work out:

 a $33\frac{1}{3}$% of 21   **b** $33\frac{1}{3}$% of 63   **c** $33\frac{1}{3}$% of 18   **d** $33\frac{1}{3}$% of 36

 e $33\frac{1}{3}$% of 42   **f** $33\frac{1}{3}$% of 330   **g** $33\frac{1}{3}$% of 600   **h** $33\frac{1}{3}$% of 84

11 Calculate:

 a 25% of 12   **b** 25% of 36   **c** 25% of 44   **d** 25% of 60

 e 20% of 15   **f** 20% of 35   **g** 20% of 40   **h** 20% of 60

12 Find:

 a 50% of 32   **b** $33^{1}/_{3}$% of 81   **c** 20% of 50   **d** 25% of 2004

 e 20% of 120   **f** 10% of 70   **g** 25% of 240   **h** $33^{1}/_{3}$% of 48

13 Ellie bought a computer. It was priced at £693.
There was a $33\frac{1}{3}$% discount.

 a How much was the discount?

 b How much did she pay?

14 Vicky sold cakes for 84p. She increased the cost by 25%.

 a What is 25% of 84?   **b** What is the new cost?

15 Dorothy was shortening curtains. They were 125 cm long.
She took 20% off their length.

 a What is 20% of 125?   **b** What is the new length of the curtains?

16 Katrona and Karen shared 450 ml of juice. Katrona took $33\frac{1}{3}$% of the juice.

 a How much juice did Katrona take?

 b How much was left for Karen?

17 Emily and Pat share the driving on a 300 mile journey.
Emily does 25% of the driving.
  a How far does Emily drive?   **b** How far does Pat drive?

18 Terri flew for 81 minutes in his microlight aircraft.
$33\frac{1}{3}$% of the time it was raining.
  a For how long was it raining?
  b For how long was it not raining?

19 Sandy, Bruce, Stephen and Alasdair shared the driving on a 600 km journey.
Sandy did $33\frac{1}{3}$% of the driving.
  a What distance did Sandy drive?
  b Bruce did 20% of the journey. How far was that?
  c Alasdair did 25% of the journey. How far did he take the car?
  d Stephen did the rest. How far did he drive?

Using a calculator to find percentages

Example 6   Find 24% of 560.

24% of 560 = 560 ÷ 100 × 24

(560) (÷) (100) (×) (24) (=) | *134·4* |

24% of 560 = 134·4

Example 7   A total of 150 fish were caught in a fishing competition.
John caught 14% of them.
How many did he catch?

(150) (÷) (100) (×) (14) (=) | *21* |

John caught 21 fish.

Exercise 3.2

1 Find:
  a 24% of 850   **b** 28% of 400   **c** 35% of 500   **d** 40% of 800
  e 26% of 700   **f** 40% of 900   **g** 35% of 200   **h** 41% of 400

2 Calculate:

   a 24% of 240   **b** 32% of 440   **c** 35% of 250   **d** 44% of 320

   e 52% of 65   **f** 65% of 82   **g** 71% of 86   **h** 88% of 150

   i 48% of 144   **j** 62% of 180   **k** 68% of 250   **l** 92% of 450

   m 72% of 44   **n** 85% of 160   **o** 95% of 450   **p** 83% of 120

3 At the start of one month, Duncan weighed 56 kg.
By the end of the month he had lost 7% of his weight.

   a How much did he lose?

   b What did he weigh at the end of the month?

4 In a race, the winner's time was 58 minutes.
The slowest athlete took 15% longer.

   a How much longer was the slowest athlete's time?

   b How long did the slowest runner take?

5 In an effort to get fit, Kirsty spends 40 minutes exercising each night.
She spends 35% of this time walking and the rest jogging.

   a How long does she spend walking?

   b How long does she spend jogging?

6 Edinburgh Castle attracted 2500 visitors one week.
64% of them were adults and the rest children.

   a How many adults were there?

   b How many children were there?

7 George won £1250 on the lottery.
He gave each of his two children 14%.

   a How much did he give each of them?

   b He then bought a television which cost £500.
How much did he have left?

8 Susan took 40 minutes to cycle from Currie into the
centre of Edinburgh.
On the way back she was cycling into a strong wind.
It took 14% longer on the way back.
How long did the whole trip take to the nearest minute?

9 Out of a year group of 180 pupils, 55% walk to school, 35% travel to school by
bus and the rest come by car.

   a How many pupils walk to school?

   b How many pupils come to school by car?

10 A garage sold 160 cars over a year.
35% of them were red, 15% were blue and the rest were silver.

    a How many of the cars sold were red?

    b How many cars were silver?

---

4 Percentage increase and decrease

Percentage increases and decreases are common in shops and businesses.

Example 1

Summer Sale
FLEECE
Original price £35
Sale 10% off marked price

    What is the sale price?

    10% of £35   = £3·50

    So sale price = £35 − £3·50

                 = £31·50

Example 2   The total rainfall for September was 360 mm.
In October, the rainfall increased by 5%.
How much rain fell in October?

    5% of 360 = 18 mm

    So rainfall = 360 + 18

               = 378 mm

Exercise 4.1

1 a Jack buys a pair of boots in the sale.
    How much do they cost?

   b Fiona buys a jacket in the sale.
    How much does it cost?

   c Alex buys an ice axe and two ropes.
    How much does he pay altogether?

Mountain Equipment Sale
Original prices
Jacket...........................£280
Ice Axe........................£90
Rope...........................£77
Boots...........................£160
Sale 10% off marked price.

2 Saima is given 8% staff discount in the shop where she works.
Her shopping bill, before her discount, is £35·50.
How much does she pay?

3 Last year Nabeel bought a bike which cost £180.
This year the same bike costs 3% more.
How much does the bike cost this year?

4 Mr and Mrs Sampson paid £475 for their car insurance last year.
This year the price has increased by 4%.
How much do they have to pay this year?

5 A school football team scored 40 goals last season.
So far this season they have scored 5% more.
How many goals have they scored so far this year?

6 Last year Mr Murray organised a school trip to Saalbach in Austria.
Each pupil had to pay £350.
This year the price for each pupil has increased by 4%.

   a How much does each pupil have to pay this year?

   b This year there are 35 pupils going on the trip.
     What is the total cost for all 35 pupils?

7 Charlotte's time for the marathon was 3 hours 20 minutes.
Michael's time was 10% slower.
What was his time?
(Hint: convert Charlotte's time to minutes.)

8 The Stuart family spent 4 hours flying from Newcastle airport to Crete.
Unfortunately, they spent 30% longer than this waiting in the airport.
How long did they have to wait in the airport?

9 Over the summer, the Millers' electricity bill was £56.
During the winter months, they expect to have a 40% increase on their
summer bill.
How much do they expect to pay for the winter months?

10 Last year, a school's maths department was given £500 to spend on jotters,
books and stationery.
This year the amount has been reduced by 5%.
How much did they get this year?

Challenge

Copy and complete the cross-number puzzle.

Across

1 50% of 52

2 $\frac{1}{5}$ of 555

4 $\frac{1}{10}$ of 750

5 $\frac{1}{7}$ of 4277

7 $\frac{1}{9}$ of 117

8 20% of 1320

9 $\frac{1}{6}$ of 1638

11 $\frac{1}{10}$ of 190

Down

1 $\frac{1}{3}$ of 738

2 $\frac{1}{8}$ of 120

3 25% of 4936

4 $\frac{1}{4}$ of 284

6 25% of 788

7 $\frac{1}{5}$ of 8355

9 70% of 40

10 $\frac{1}{9}$ of 351

5 Rates revisited

Example 1 Grace types at the rate of 60 words per minute.
This means she can type 60 words in 1 minute.
How many words can she type in 4 minutes at this rate?

$$4 \times 60 = 240 \text{ words}$$

Example 2 A car travels at 30 miles per hour.
How far will it travel in 4 hours?

In 1 hour the car will travel 30 miles.
So in 4 hours it will travel $4 \times 30 = 120$ miles.

Exercise 5.1

1 It costs £3 for one adult to go to the cinema.
How much will it cost for eight adults?

2 A car is travelling at 70 miles per hour.
How far will it travel in 5 hours at this speed?

3 Sam can walk at 6 kilometres per hour.
At this speed, how far will he walk in 8 hours?

4 Mr Robertson paid £350 per adult and £99 per child for a two week holiday.

  a How much did he pay for his three children?

  b How much did he pay altogether for two adults and three children?

5 Ryan is paid £5·60 per hour in his job.
How much is he paid for working an 8 hour day?

6 At birth, a baby great white shark is 1·5 metres in length.
It can grow at a rate of 25 centimetres per year.
How long will it be after 8 years?

7 A swallow can fly at a speed of 20 miles per hour.

  a How far could it fly in a day at this speed?

  b How far could it fly in 30 days?

8 At Edinburgh airport, planes are either taking off or landing at a rate of two every minute.
How many planes will either take off or land in 30 minutes?

Working out rates

Example   Six bags of crisps can be bought for £1·80.
Calculate the cost per bag of crisps.

The word 'per' is a clue to divide, in this case by the number of bags.
So the cost per bag = £1·80 ÷ 6 = £0·30

Exercise 5.2

1 Jo types 300 words in 6 minutes.
Calculate the rate at which she types (in words per minute).

2 Two hundred and forty cars passed a checkpoint in 8 minutes.
Calculate the rate at which they passed the checkpoint (in cars per minute).

3 A lemonade company can fill 500 bottles in 4 minutes.
Calculate the rate at which the bottles are filled
(in bottles per minute).

4 After exercising, Kevin counted 360 heart beats in 3 minutes.
Calculate the rate at which his heart was beating (in beats per minute).

5 In 8 weeks, Sammy lost 36 pounds in weight.
Calculate the rate at which he lost weight (in pounds per week).

6 In training, Kirsty ran 340 miles in 4 weeks.
 a Calculate her running rate in miles per week.
 b Calculate her running rate in miles per day.

◀◀ **RECAP**

Fractions
You must be able to calculate fractions of quantities, for example:

$\frac{1}{3}$ of $96 = 96 \div 3 = 32$

Percentages
You must be able to write percentages as fractions and as decimals, for example:

$30\% = \frac{30}{100} = 0{\cdot}3$

You must be able to calculate a percentage of an amount without a calculator.
To do this you should use common fractions instead of percentages.
You should know that:

$50\% = \frac{1}{2}$
$33\frac{1}{3}\% = \frac{1}{3}$
$25\% = \frac{1}{4}$
$20\% = \frac{1}{5}$
$10\% = \frac{1}{10}$
$1\% = \frac{1}{100}$

You must be able to calculate a whole number percentage of an amount using a calculator.
For example, to find 28% of 300 enter:

$\boxed{300}\ \boxed{\div}\ \boxed{100}\ \boxed{\times}\ \boxed{28}\ \boxed{=}\qquad \boxed{\textit{84}}$

So 28% of 300 = 84

Rates
You must know how to solve problems using rates.

Example A car is using petrol at a rate of 40 miles per gallon.
How far will it go on 6 gallons of petrol?

Answer $6 \times 40 = 240$ miles

---

1 Find:
   a $\frac{1}{2}$ of 76   **b** $\frac{1}{3}$ of 69   **c** $\frac{1}{5}$ of 85   **d** $\frac{1}{6}$ of 126
   e $\frac{1}{3}$ of 360   **f** $\frac{1}{4}$ of 600   **g** $\frac{1}{8}$ of 128   **h** $\frac{1}{7}$ of 420
   i $\frac{1}{9}$ of £450   **j** $\frac{1}{6}$ of 432   **k** $\frac{1}{7}$ of 6454   **l** $\frac{1}{8}$ of 6644

2 Calculate:
   a $\frac{1}{2} \times 1{\cdot}24$ **b** $\frac{1}{3} \times 6{\cdot}24$ **c** $\frac{1}{5} \times 4{\cdot}65$ **d** $\frac{1}{4} \times 5{\cdot}40$
   e $\frac{1}{6} \times 25{\cdot}32$ **f** $\frac{1}{5} \times 20{\cdot}45$ **g** $\frac{1}{3} \times 47{\cdot}04$ **h** $\frac{1}{6} \times 70{\cdot}02$

3 Work out:
   a 50% of 480   **b** 25% of 680   **c** 20% of 480   **d** $33\frac{1}{3}$% of 510

4 Use your calculator to find:
   a 42% of 320   **b** 66% of 860   **c** 72% of 550   **d** 45% of 900

5 Last year Lauren paid £7800 for her new car.
   This year the same car costs 10% less.
   How much does the car cost this year?

6 Stephen took 24 minutes to canoe down a stretch of river.
   On the return trip it took him $33\frac{1}{3}$% longer.
   How long did the whole trip take?

7 It costs £15·36 to hire an indoor tennis court.
   An outdoor court costs $\frac{1}{3}$ less.
   How much does it cost for an outdoor court?

8 There are 50 litres of water in a barrel.
   Water is dripping out of it at a rate of 2 litres per minute.
   How long will it take to empty?

9 Shona can do sit-ups at a rate of 36 per minute.
   How many will she be able to do in two and a half minutes?

10 Charlotte collected 15 bags of rubbish in 45 minutes.
   Calculate the rate at which she collected rubbish in minutes per bag.

6 Scale drawing

In 1492, Christopher Columbus used maps to navigate.

To find his new position, he measured the distance and direction from his old position.

The direction was worked out using a compass.

The distance was worked out by figuring for how long a certain speed was kept up.

1 Review

◀◀ Exercise 1.1

1 Measure the length of each snail trail.

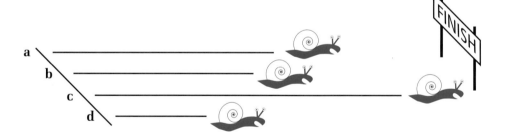

2 Draw lines that measure:

    a 5 cm   **b** 3·2 cm   **c** 23 mm   **d** 44 mm

3 Convert these distances into metres:

    a 500 cm   **b** 380 cm   **c** 624 cm   **d** 157 cm

4 Convert these distances into centimetres:

    a 4 m   **b** 2·6 m   **c** 1·92 m   **d** 3·84 m

5 Convert these distances into metres:

    a 6 km   **b** 3·4 km   **c** 8·268 km   **d** 1·045 km   **e** 0·42 km

6 Convert these distances into kilometres:

    a 8000 m   **b** 7200 m   **c** 6390 m   **d** 2145 m   **e** 360 m

7 Measure each angle.

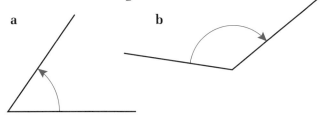

8 Draw an angle of size:

    a 30°   **b** 72°   **c** 135°   **d** 177°

9 Make an accurate drawing of this triangle.

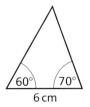

10 a What hill is south of Tom?

    b What hill is west of Tom?

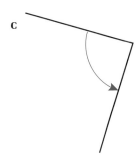

11 Using the map of the park, write down the coordinates of:

    a the tree house

    b the swings

    c the pond

    d the marsh.

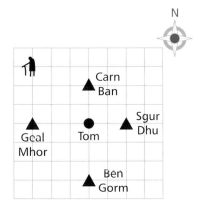

2 The eight point compass

The eight main directions of the compass are called
the cardinal points.

North, south, east and west are well known.

The other four lie *halfway* between these points and
take their names from them.

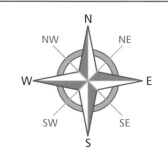

Example

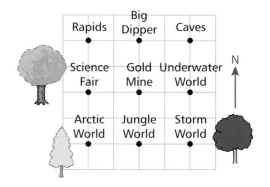

Ben Nevis lies north-west (NW) of Binnean Mor.

Binnean Mor lies south-east (SE) of Ben Nevis.

Exercise 2.1

1 This tourist map shows the layout of an
   adventure park.

   a What lies NE of the Gold Mine?

   b What lies SW of Underwater World?

   c What lies SE of the Rapids?

   d If I walk NW from Underwater World,
      where would I arrive?

   e If I walk NE from the Science Fair, where
      would I arrive?

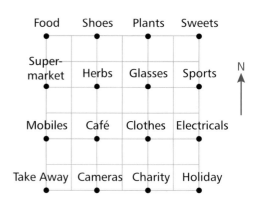

2 This diagram shows the different types of
   shops in a shopping centre.

   a What shop lies NE of the herb shop?

   b What shops lie SW of the sports shop?

   c What shop lies SW of the café?

   d What shop lies SE of the plant shop?

   e I walk NE from the mobile phone shop.
      What is the next shop I will arrive at?

   f I walk SW from the sweet shop.
      What is the second shop I will arrive at?

   g I walk from the food shop to the herb shop.
      From there, I walk to the supermarket and on to the café.
      Write down the directions I followed.

3 a I sailed south. What direction will I need to follow to get back to where I started?

   b I walked SW. What direction will I need to take to get back to where I started?

   c I walked NW. What direction will I need to follow to get back to where I started?

4 a What lies NW of the car park?

   b What lies SE of Green Hill?

   c I walked from the car park to Green Hill. Then I walked to the viewpoint and on to the forest before finally walking back to the car park. What directions did I follow?

   d I walked following these directions: S, then SE. I ended up in the car park. At what named point did I start?

   e I walked N, NW, S and then E, ending up in the forest. I always changed direction at a named point. Where did I start?

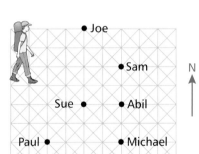

5 a Sue is facing Joe.

   i She turns clockwise to face Abil. How many degrees has she turned through?

   ii How many degrees would she have turned through if she had turned to Sam instead?

   b Sue faces Sam. She turns clockwise to face Paul. How many degrees has she turned through?

   c Sue faces Sam. She turns clockwise to face Michael. How many degrees has she turned through?

   d Sue faces Michael. She turns anticlockwise to face Joe. How many degrees has she turned through?

6 How many degrees are there between:

   a NW and N   **b** SE and SW   **c** S and NW   **d** NW and SE?

3 Bearings

Mountaineers use bearings to navigate safely in the hills.

A **bearing** is an angle, measured **clockwise** from **north**.
All bearings are given as **3 figures**.
94° becomes 094°.
7° becomes 007°.

Example 1 This hiker is walking on a bearing of 120°.

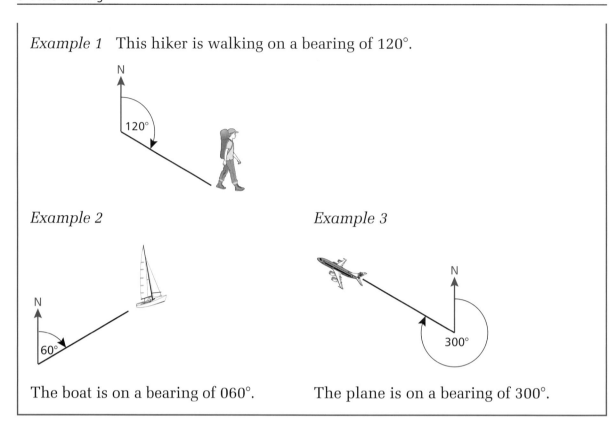

Example 2

Example 3

The boat is on a bearing of 060°.

The plane is on a bearing of 300°.

Exercise 3.1

1 Six vessels are spotted from a lighthouse, L.
Measure the bearing of each vessel from the lighthouse.

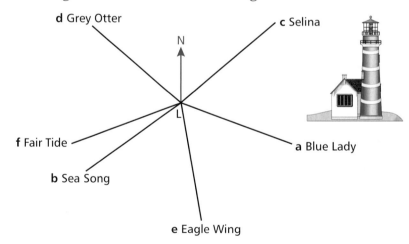

2 Draw each bearing below. Remember to measure the angle clockwise from north.

    a 045°   **b** 110°   **c** 165°   **d** 205°

    e 340°   **f** 280°   **g** 190°   **h** 005°

3 Write down the bearings from north to:

    a east   **b** west   **c** north-east   **d** south-west

4 If you were walking on these bearings, what compass direction would you be following?

   a 135°   **b** 180°   **c** 000°   **d** 315°

5 Measure the bearings of each leg in this orienteering course.

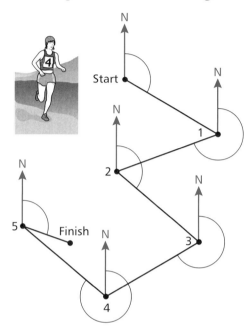

4 Scales

In order to make sense of any map or scale diagram, the scale must be known.

Here is a scale drawing of a car.
The scale on the diagram tells us:
1 cm represents 1 metre.
This means that for every 1 metre of the actual car,
1 cm is drawn on paper.
The actual car is 4 metres in length,
so the scale drawing is 4 cm.

1 cm represents 1 m

Example   The height of the church in the
               drawing is 3 cm.
               What is the real height of the church?

               1 cm represents 20 metres.
               So 3 cm represents $3 \times 20 = 60$ metres.

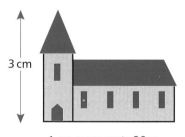

1 cm represents 20 m

Exercise 4.1

1 For each diagram:
   i measure the height   **ii** use the scale to calculate the height in real life.

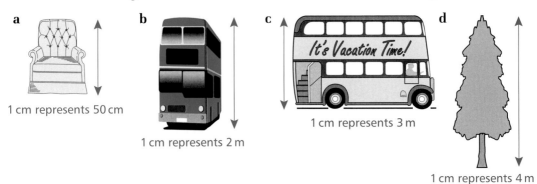

a

1 cm represents 50 cm

b

1 cm represents 2 m

c

It's Vacation Time!

1 cm represents 3 m

d

1 cm represents 4 m

2 The map shows the route of a cross-country course.

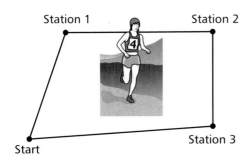

Station 1   Station 2

Start   Station 3

1 cm represents 100 metres

  a Measure the length of each section of the course.
  b Work out the actual distance of the whole course.

3 Copy and complete the table.

	Scale of map	Actual distance	Distance on map
a	1 cm represents 200 m	400 m	
b	1 cm represents 200 m	100 m	
c	1 cm represents 500 m	1 km	
d	1 cm represents 2 km	4 km	
e	1 cm represents 3 km	12 km	
f	1 cm represents 50 km	400 km	
g	1 cm represents 100 km	250 km	
h	1 cm represents 20 m		4 cm
i	1 cm represents 500 m		6 cm
j	1 cm represents 10 km		3·5 cm

4 The ship, the *Sunstar*, sailed from port to Lighthouse Point.
It then sailed on to Gull Point before returning to port.

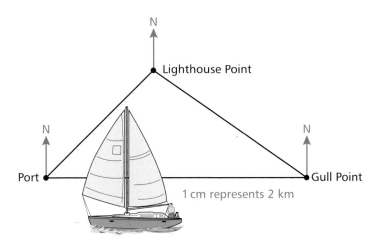

1 cm represents 2 km

Copy and complete the table.

	Bearing followed	Distance on map	Actual distance
Port to Lighthouse Point			
Lighthouse Point to Gull Point			
Gull Point to port			

5 This is a map of an adventure park.

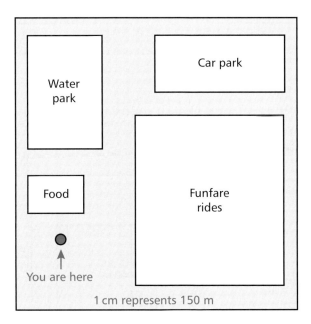

1 cm represents 150 m

a Measure the dimensions of the four named areas of the park.
b Work out the real size of these areas.

5 Making scale drawings

Scale drawings should be as large as possible if measurements are to be taken from them.
Make a rough sketch of the situation to help you plan your steps.

Example An architect has drawn a rough sketch of one wall of a hall. She needs to make an accurate scale drawing of the wall.

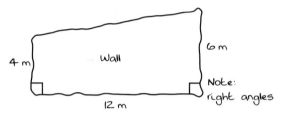

First she has to choose a suitable scale.
She decides to let 1 cm represent 2 metres.
So 12 m becomes 6 cm on the drawing,
6 m becomes 3 cm and 4 m becomes 2 cm.

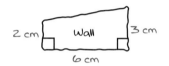

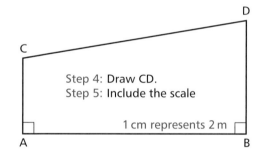

Step 1: Draw a 6 cm line, AB.
Step 2: Draw a 2 cm line, AC, at right angles to AB.
Step 3: Draw a 3 cm line, BD, at right angles to AB.

Step 4: Draw CD.
Step 5: Include the scale

1 cm represents 2 m

Exercise 5.1

1 Make accurate drawings of each sketch using the scales given.
Mark the lengths on each side.
Write the scale beside each diagram.

a

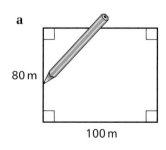

80 m

100 m

1 cm represents 10 metres

b

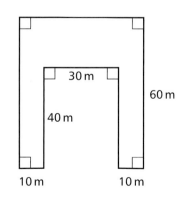

30 m

60 m

40 m

10 m   10 m

1 cm represents 10 metres

c

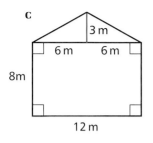

1 cm represents 2 metres

d

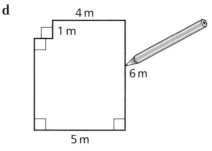

4 m
1 m
6 m
5 m

1 cm represents 50 cm

e

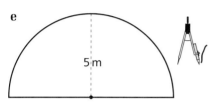

5 m

1 cm represents 2 metres

f

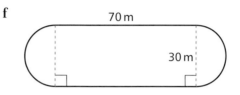

70 m
30 m

1 cm represents 10 metres

6 More scale drawings

Remember: When drawing bearings, draw a north line.
Measure the bearing from north in a clockwise direction.

Example The *Jupiter* sailed on a bearing of 030° for 50 km.
Make an accurate scale drawing of the ship's route using 1 cm to represent 10 km.

Step 1: Make a sketch of the situation.

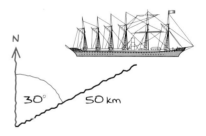

Step 2: Draw a north line.

Step 3: Measure an angle of 30° from north, clockwise.

Step 4: Using the scale, 50 km will be represented by 5 cm.
Draw a 5 cm line in the right direction.

Step 5: Include the scale on the drawing.

Jupiter
N
50 km
30°

1 cm represents 10 km

Exercise 6.1

1 Make accurate scale drawings of these sketches using the given scales.

a
80°
12 km

1 cm represents 4 km

b
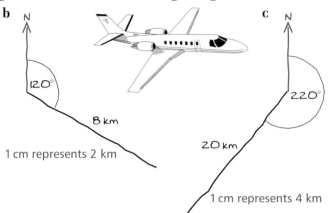
120°
8 km

1 cm represents 2 km

c
220°
20 km

1 cm represents 4 km

2 These journeys have two legs.
Make an accurate scale drawing of each journey using the given scales.

a
150°
40°
8 km
10 km

1 cm represents 2 km

b
1 cm represents 3 km
170°
12 km
9 km
20°

3 Ben was taking part in a hill race. First he ran 3 km south.
Then he ran 5 km on a bearing of 045° to the finish.
a Make an accurate scale drawing of his route.
(Use 1 cm to represent 1 km.)
b Measure the direct distance, on your diagram, from start to finish.
c What is the actual distance represented by this map distance?

4 The *Beartha* sailed on a bearing of 050° for 6 km.
It then sailed for a further 5 km on a bearing of 160°.
a Using 1 cm to represent 1 km, make an accurate scale drawing of the boat's route.
b Measure the distance between the starting and finishing point.
c What is the actual distance between start and finish?

5 Emma cycled 6 km from her house on a bearing of 200°.
She then cycled a further 8 km due west to the beach.
a Using 1 cm to represent 2 km, make an accurate scale
drawing of her route.
b From the beach, she cycled directly back home.
How far did she have to cycle, to the nearest kilometre?

6 Two planes left Edinburgh airport.
   One flew on a bearing of 170° for 200 km.
   The other flew on a bearing of 240° for 400 km.
   a Using 1 cm to represent 100 km, make an accurate
   drawing showing both routes on the one diagram.
   b How far apart are the two destinations?

7 Directions on maps

Maps can be used to give directions.

Example   Cameron is in the shoe shop.
How does he get to the bank?

He turns left out of the shoe shop
onto Heather Drive.
He takes the first road on his right,
Beach Road.
After passing Tulip Street, the bank is
on his right.

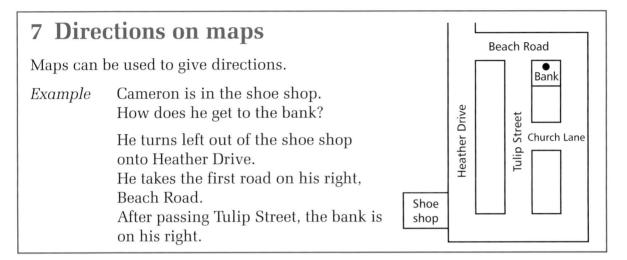

Exercise 7.1

Use this map to answer questions **1** to **5**.

1 Jamila is standing at A facing west.
   a What road is she on?
   b What road is first on her left?
   c What road is second on her left?
   d What road is first on her right?
   e What road is second on her right?

2 Lewis is standing at B looking south.
   a What road is he on?
   b What is the first road on his right?
   c He takes the first road on his right
      and stops.
      i What shop is on his right?
      ii What road is next on his left?

3 Zoe is in the park.
   She walks along Twig Lane, turns
   left then first right.
   a What road is she now on?
   b What road is next on her right?

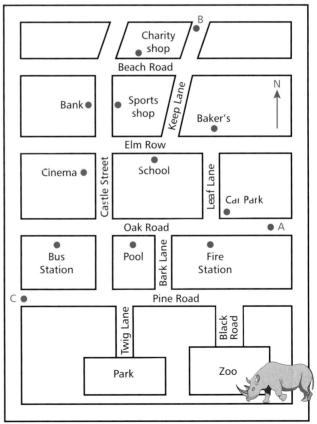

4 Zac is standing at C looking east.

   a He takes the first road on his left.
      Then he takes the first right and first left.
      Write down the roads he has walked along.

   b Back at C, Zac now takes the second road on his left.
      Then he turns left and first right.
      What road is he now on?

   c From C, what road will Zac be on if he takes the first left and second right?

   d Give directions for Zac to get from C to:
      i the bank   **ii** the zoo   **iii** the charity shop   **iv** the baker's.

5 Give directions to get from:

   a the baker's to the zoo

   b the cinema to the charity shop

   c the park to the sports shop.

---

8 Enlarging and reducing

These two photos are of the same view.

2 cm

1·5 cm

3 cm

4 cm

The second photo is double the size of the first photo – the first could be **enlarged** to make the second.
Each length in the second photo is double the size of that in the first.

We could say that the first photo is half the size of the second one – the second could be **reduced** to make the first.

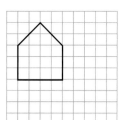

This shape is enlarged by counting squares and doubling all the lengths.

Double the horizontal and vertical lines – the sloping ones should take care of themselves.

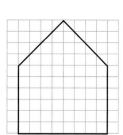

Exercise 8.1

1 On squared paper copy each diagram and then draw one twice as big.

a 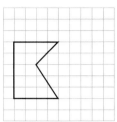 **b** 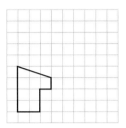 **c**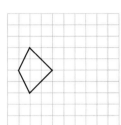

(Hint: draw the diagonals.)

2 On squared paper copy each diagram and then draw one half the size.

a 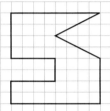 **b** 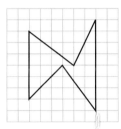 **c**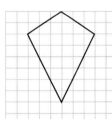

(Hint: draw the diagonals.)

9 Similar rectangles

Two rectangles are **similar** if one rectangle is an enlargement of the other.

Example
These two rectangles are similar.
The second rectangle is double
the size of the other.
It has been enlarged by a **factor** of 2.

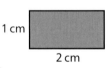 

Missing lengths can be worked out
if we know the rectangles are similar.

Example
These two rectangles are similar.
Work out the height of the
larger one.

 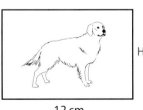

The width has enlarged from 3 cm to 12 cm.
It's 4 times bigger.
(It has been enlarged by a factor of 4.)
So the height must also be 4 times bigger: $2 \times 4 = 8$.
The larger rectangle has a height of 8 cm.

Exercise 9.1

1 By what factor has each rectangle been enlarged?

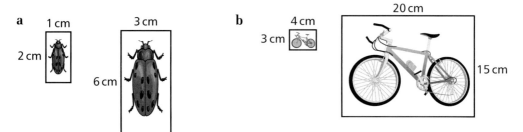

2 There are two rectangular warning signs beside a swimming pool. They are similar.

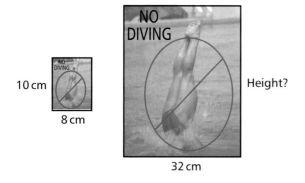

a What is the enlargement factor?

b What is the height of the larger sign?

3 The lids of two rectangular chocolate boxes are similar.

a What is the enlargement factor?

b What is the width of the larger lid?

4 Each pair of rectangles is similar. Calculate the missing lengths.

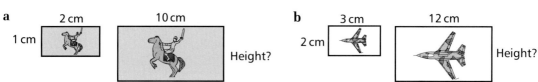

5 Each pair of rectangles is similar.

a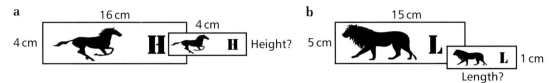

b

The rectangles have been shrunk, one by a factor of 4, the other by a factor of 5.
Calculate the missing lengths.

6 A door in a doll's house is similar to a real size door.
The doll's house door is 6 cm wide and 10 cm in height.
The real door has a height of 2 metres.
Calculate the width of a real door.

◀◀ RECAP

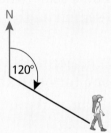

The eight point compass
You need to know the names of the eight points and
how to use the compass.

Bearings
Bearings are 3-digit numbers.
They describe direction.
They are angles, always measured
from north in a clockwise direction.

Scales and scale drawings
Scales are used to construct scale drawings, maps and plans.
Scale drawings should be as large as possible if measurements are to be taken
from them.
A rough sketch of the situation should be made to help you plan your steps.

Directions
You need to be able to understand and give directions relating to a map.

Enlarging and reducing
You must be able to draw shapes double or half their size.

Similar rectangles
If you know rectangles are similar you should be able to:
● work out the scale factor
● calculate missing lengths.

1 The diagram shows the layout of tents at a scout camp.
Which tent is:

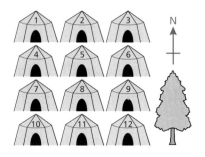

a north of tent 5

b south-west of tent 8

c north-east of tent 8

d south-east of tent 4?

2 Measure each bearing.

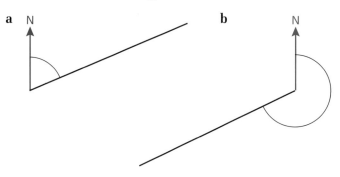

3 Adam and his friends walked for 10 km.
The map they were using had a scale where 1 cm represented 2 km.
How would 10 km be represented on the map?

4 The *Isaac Newton* sailed on a bearing of 050° for 10 km.
Then it sailed south-east for a further 6 km.

a Using 1 cm to represent 2 km, make an accurate
scale drawing of the ship's route.

b Measure the distance between the starting point
and the finishing point of the trip.

c How far has the ship to sail to get back to where
it started?

5 a Susan is at the police station.
Write directions for Susan to get back
to her house.

b Tom turns left out of the butcher's shop.
He then takes the first left and
second left.
He goes into the shop on his left.
What shop is he in?

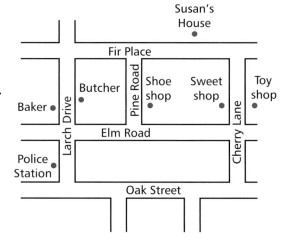

REVISE

6 Draw this shape twice as large.

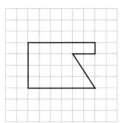

7 These two television sets are similar.
Calculate the missing height.

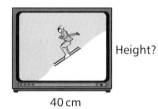

24 cm

30 cm

Height?

40 cm

7 Patterns and formulae

A formula is like a recipe.
It gives the instructions for a calculation.

The formula for calculating calories burnt is:
number of calories burnt
= (body weight in kilograms) × (kilometres run)

A 60 kg runner in a 40 km race burns
60 × 40 = 2400 calories.

1 Review

◀◀ Exercise 1.1

1 Draw the next two designs in each pattern.

a

1   2   3   ?   ?
                          4   5

b

1   2   3   ?   ?
                          4   5

2 Here are some patterns and some rules.
The rules tell you how to get the next term from the one before it.
For example, 3, 7, 11, 15, ... has the rule add 4.
Match these number patterns and rules.

a

Patterns				
A:	2	6	10	14	...
B:	1	3	5	7	...
C:	10	9	8	7	...

Rules	
1:	Add 2
2:	Take away 1
3:	Add 4

b

Patterns				
D:	3	7	11	15	...
E:	3	8	13	17	...
F:	3	5	7	9	...

Rules	
4:	Add 2
5:	Add 4
6:	Add 5

3 Find the next three numbers in each simple pattern:

    a 7, 6, 5, 4, …   **b** 3, 5, 7, 9, …   **c** 1, 5, 9, 13, …

    d 23, 20, 17, 14, …   **e** 55, 50, 45, 40, …   **f** 32, 28, 24, 20, ..

4 Copy and complete this table:

Number of skateboards	1	2	3	4
Number of wheels	4	?	?	?

5 Use the rule to help you work out the next two numbers in each pattern.

    a   Double: 1 — 2 — 4 — ? — ?

    b   Halve: 32 — 16 — 8 — ? — ?

---

2 Picture and number patterns

Number patterns have rules for turning one term into the next.

Example 1   An adding rule: | + 2 > | 1 | 3 | 5 | 7 | 9 | 11 | ⟹

Example 2   A subtracting rule: | − 5 > | 30 | 25 | 20 | 15 | 10 | 5 | ⟹

Example 3   A multiplying rule: | × 2 > | 1 | 2 | 4 | 8 | 16 | 32 | ⟹

Example 4   A dividing rule: | ÷ 2 > | 96 | 48 | 24 | 12 | 6 | 3 | ⟹

If you know the rule and some of the pattern, you can continue it.

The rule is:	The pattern is:	The next 2 terms are:
+ 7 >	4, 11, 18, 25, ⟹	32, 39

Exercise 2.1

1 Find the missing terms for each pattern.

  a | + 7 > | 2 | 9 | 16 | ? | ? | ? | ⟹   **b** | − 3 > | 20 | 17 | 14 | ? | ? | ? | ⟹

  c | × 2 > | 3 | 6 | 12 | ? | ? | ? | ⟹   **d** | + 2 > | 5 | 7 | 9 | ? | ? | ? | ⟹

  e | + 4 > | 10 | 14 | 18 | ? | ? | ? | ⟹   **f** | − 5 > | 28 | 23 | 18 | ? | ? | ? | ⟹

  g | ÷ 2 > | 64 | 32 | 16 | ? | ? | ? | ⟹   **h** | − 10 > | 90 | 80 | 70 | ? | ? | ? | ⟹

2 Match each pattern to a rule.

a

Patterns		Rules	
A:	10, 11, 12, 13, …	1:	Add 2
B:	10, 8, 6, 4	2:	Subtract 1
C:	10, 12, 14, 16, …	3:	Take away 2
D:	10, 9, 8, 7, …	4:	Add 1

b

Patterns		Rules	
A:	40, 20, 10, 5, …	1:	Multiply by 2
B:	40, 42, 44, 46, …	2:	Divide by 2
C:	40, 80, 160, 320, …	3:	Add 2
D:	40, 38, 36, 34, …	4:	Subtract 2

3 i Draw the next two missing pictures. **ii** Find the next two missing terms.

a

$$4 \qquad 6 \qquad 8$$

b

$$5 \qquad 8 \qquad 11$$

c

$$6 \qquad 8 \qquad 10$$

d

$$9 \qquad 13 \qquad 17$$

4 For each number pattern find:
  i the rule for getting the next term   **ii** the missing terms.
  a 3, 7, 11, 15, ?, ?   **b** 33, 29, 25, 21, ?, ?   **c** 8, 14, 20, 26, ?, ?
  d 1, 8, 15, 22, ?, ?   **e** 2, 4, 8, 16, ?, ?   **f** 243, 81, 27, 9, ?, ?

Challenges

1 Two four-card patterns have been made, one using the rule 'Add 3' and the other using the rule 'Take away 2'.
The cards have then been mixed together and arranged in order.

Can you sort the cards back out into the two patterns?

2 Three four-card patterns have been made, one using the rule 'Add 2', a second using the rule 'Take away 2' and a third using 'Times 2'.
The cards have then been mixed together and arranged in order.

Can you sort the cards back into the three patterns?

---

3 New patterns from old

The simple pattern 1, 2, 3, 4, 5, … uses the 'counting numbers'.
We can make up rules to turn the counting numbers into other patterns.

Example 1   What pattern do you get when you add 5 to the counting numbers?

Counting numbers	1	2	3	4	5	…
Patterns	6	7	8	9	10	…

Rule
+ 5

You get: 6, 7, 8, 9, 10, …

Example 2   What pattern do you get when you double the counting numbers?

Counting numbers	1	2	3	4	5	…
Patterns	2	4	6	8	10	…

Rule
× 2

You get: 2, 4, 6, 8, 10, …

Exercise 3.1

1 Work out the new pattern in each case.

a

Counting numbers	1	2	3	4	...
Pattern					...

Rule +3

b

Counting numbers	1	2	3	4	...
Pattern					...

Rule ×4

c

Counting numbers	1	2	3	4	...
Pattern					...

Rule +10

d

Counting numbers	1	2	3	4	...
Pattern					...

Rule ×7

2 Find the missing rule in each case.

a

Counting numbers	1	2	3	4	...
Pattern	3	6	9	12	...

Rule

b

Counting numbers	1	2	3	4	...
Pattern	6	7	8	9	...

Rule

c

Counting numbers	1	2	3	4	...
Pattern	0	1	2	3	...

Rule

d

Counting numbers	1	2	3	4	...
Pattern	8	16	24	32	...

Rule

3 Find the rule which changes:

a 1, 2, 3, 4, ... into 9, 10, 11, 12, ...   **b** 1, 2, 3, 4, ... into 4, 5, 6, 7, ...

c 3, 4, 5, 6, ... into 1, 2, 3, 4, ...   **d** 6, 7, 8, 9, ... into 2, 3, 4, 5, ...

e 1, 2, 3, 4, ... into 7, 14, 21, 28, ...   **f** 1, 2, 3, 4, ... into 11, 22, 33, 44, ...

g 2, 4, 6, 8, ... into 1, 2, 3, 4, ...   **h** 8, 12, 16, 20, ... into 2, 3, 4, 5, ...

i 1, 2, 3, 4, ... into 2, 4, 6, 8, ...   **j** 1, 2, 3, 4, .. into 6, 12, 18, 24, ...

k 1, 2, 3, 4, ... into 9, 18, 27, 36, ...   **l** 1, 2, 3, 4, ... into 5, 10, 15, 20, ...

4 Copy and complete these tables:

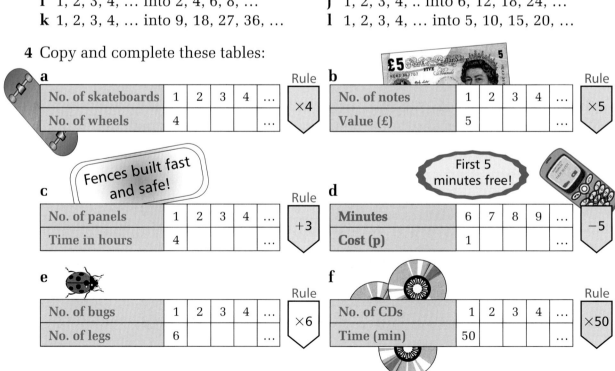

a

No. of skateboards	1	2	3	4	...
No. of wheels	4				...

Rule ×4

b

No. of notes	1	2	3	4	...
Value (£)	5				...

Rule ×5

Fences built fast and safe!

c

No. of panels	1	2	3	4	...
Time in hours	4				...

Rule +3

First 5 minutes free!

d

Minutes	6	7	8	9	...
Cost (p)	1				...

Rule −5

e

No. of bugs	1	2	3	4	...
No. of legs	6				...

Rule ×6

f

No. of CDs	1	2	3	4	...
Time (min)	50				...

Rule ×50

5 Find the missing entries using the rule given.

a

Counting numbers	1	2	3	4		16
Pattern						...

Rule +7

b

Counting numbers	4	5	6	7		23
Pattern						...

Rule −3

c

Counting numbers	1	2	3	4		9
Pattern						...

Rule ×20

d

Counting numbers	1	2	3	4		20
Pattern						...

Rule ×9

e

Counting numbers	7	8	9	10		33
Pattern						...

Rule −5

f

Counting numbers	1	2	3	4		18
Pattern						...

Rule +12

Challenge

Obey the rules!
Copy and complete each table.

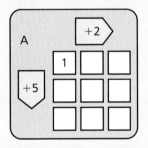

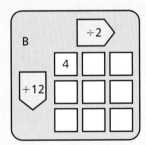

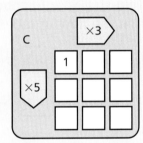

Two-step rules
Some patterns have two-step rules.
Look at how this fence is built from posts and crossbars.

No. of posts	2	3	4	5
No. of crossbars	3	6	9	12

The first row becomes the second in two steps:

No. of posts	2	3	4	5
Step 1: ×3	6	9	12	15
Step 2: −3	3	6	9	12

To find the number of crossbars multiply the number of posts by 3 then take away 3.

Exercise 3.2

1 Copy and complete each table.

a

Counting numbers	1	2	3	4	5
Step 1: ×2					
Step 2: +3					

b

Counting numbers	1	2	3	4	5
Step 1: ×7					
Step 2: −5					

c

Counting numbers	1	2	3	4	5
Step 1: ×3					
Step 2: −1					

d

Counting numbers	1	2	3	4	5
Step 1: ×10					
Step 2: +6					

e

Counting numbers	1	2	3	4	5
Step 1: ×4					
Step 2: +7					

f

Counting numbers	1	2	3	4	5
Step 1: ×9					
Step 2: −8					

2 Use the two-step rule to change the pattern.
Write down the new number pattern in each case.

a 1, 2, 3, 4, 5, ...   — ×5 > — 3 >—

b 4, 5, 6, 7, 8, ...   — ×2 > — 7 >—

c 1, 2, 3, 4, 5, ...   — ×3 > + 6 >—

d 1, 2, 3, 4, 5, ...   — ×6 > — 2 >—

e 4, 5, 6, 7, 8, ...   — ×4 > + 1 >—

f 1, 2, 3, 4, 5, ...   — ×8 > — 2 >—

3 Match the two-step rules with the new patterns.

Rule A   **Rule B**   **Rule C**

— ×2 > + 5 >—   — ×3 > − 1 >—   — ×2 > − 1 >—

Starting with 1, 2, 3, 4, 5, ... we end with:

Pattern 1   **Pattern 2**   **Pattern 3**

2, 5, 8, 11, 14, ...   1, 3, 5, 7, 9, ...   7, 9, 11, 13, 15, ...

4 Find the missing numbers for these dot and match patterns.

a

Pattern:					
Pattern number:	1	2	3	4	5
	2				
Dots:	3				

× 2
+ 1

b

Pattern:					
Pattern number:	1	2	3	4	5
	3				
Matches:	4				

× 3
+ 1

c

Pattern:					
Pattern number:	1	2	3	4	5
	3				
Dots:	2				

× 3
− 1

d

Pattern:					
Pattern number:	1	2	3	4	5
	4				
Matches:	3				

× 4
− 1

e

Pattern:					
Pattern number:	1	2	3	4	5
	4				
Dots:	5				

× 4
+ 1

5 a It costs £25 to hire a boat and then a further £10 for each hour it is used.
Copy and complete the table to show the costs.

Hours	1	2	3	4	5
Step 1: ×10	10				
Step 2: +25: Cost (£)	35				

Only £25 and £10 for each hour

BOAT HIRE

Note that the cost in pounds = 10 times the number of hours plus 25.

b It costs £10 to hire a bike plus £5 for each hour it is used.
Copy and complete the table.

Hours	1	2	3	4	5
Step 1: ×5	5				
Step 2: +10: Cost (£)	15				

£10 charge then only £5 per hour

BLAZING SADDLES BIKE HIRE

c It costs £100 to hire a recording studio and an
additional £50 for each hour it is used.
Copy and complete the table.

Hours	1	2	3	4	5
Step 1: ×50	50				
Step 2: +100: Cost (£)	150				

Flat fee £100 then £50 per hour

4 What's my rule?

Example   How can I turn the numbers 1, 2, 3, 4, 5, … into the pattern 5, 8, 11, 14, 17, … using a two-step rule?

Step 1   We see that the pattern goes up in threes … like the '3 times' table!

     5 (+3) 8 (+3) 11 (+3) 14 (+3) 17 (+3) …

This suggests that the first step is ⟩× 3⟩

Step 2   Compare the '3 times' table with the pattern.

Counting numbers	1	2	3	4	5	…
×3	3	6	9	12	15	…
+? Pattern	5	8	11	14	17	…

It is easy to see that step 2 must be ⟩+ 2⟩

So the rule is: take a counting number, multiply by 3 then add 2.

Counting numbers —⟩× 3⟩—⟩+ 2⟩— Pattern

Exercise 4.1

1 Complete each table to help you find the two-step rule for each pattern.

a

Counting numbers	1	2	3	4	5	...
Step 1: ×?						...
Step 2: −? Pattern	3	7	11	15	19	...

b

Counting numbers	1	2	3	4	5	...
Step 1: ×?						...
Step 2: +? Pattern	7	10	13	16	19	...

c

Counting numbers	1	2	3	4	5	...
Step 1: ×?						...
Step 2: +? Pattern	9	11	13	15	17	...

d

Counting numbers	1	2	3	4	5	...
Step 1: ×?						...
Step 2: −? Pattern	3	8	13	18	23	...

e

Counting numbers	1	2	3	4	5	...
Step 1: ×?						...
Step 2: −? Pattern	2	5	8	11	14	...

f

Counting numbers	1	2	3	4	5	...
Step 1: ×?						...
Step 2: +? Pattern	7	13	19	25	31	...

2 Start with 1, 2, 3, 4, 5, ...

Find the one-step $\boxed{\times\,?}$ rule that turns the pattern into:

a 4, 8, 12, 16, 20, ...   **b** 3, 6, 9, 12, 15, ...

c 7, 14, 21, 28, 35, ...   **d** 2, 4, 6, 8, 10, ...

e 10, 20, 30, 40, 50, ...   **f** 5, 10, 15, 20, 25, ...

3 Start with 1, 2, 3, 4, 5, ...

Find the two-step rule $\boxed{\times\,?} \;\; \boxed{+\,?}$ that turns it into:

a 5, 7, 9, 11, 13, ...   **b** 4, 7, 10, 13, 16, ...

c 8, 15, 22, 29, 36, ...   **d** 9, 11, 13, 15, 17, ...

e 10, 13, 16, 19, 22, ...   **f** 9, 17, 25, 33, 41, ...

g 10, 18, 26, 34, 42, ...   **h** 12, 23, 34, 45, 56, ...

i 9, 16, 23, 30, 37, ...

4 Start with 1, 2, 3, 4, 5, ...

Find the two-step $\boxed{\times\,?} \;\; \boxed{-\,?}$ rule that turns it into:

a 1, 3, 5, 7, 9, ...   **b** 5, 11, 17, 23, 29, ...

c 1, 4, 7, 10, 13, ...   **d** 2, 7, 12, 17, 22, ...

e 4, 11, 18, 25, 32, ...   **f** 1, 11, 21, 31, 41, ...

g 5, 12, 19, 26, 33, ...   **h** 1, 5, 9, 13, 17, ...

i 1, 21, 41, 61, 81, ...

5 In each case write down the two-step rule in words.
For example: 'to find the cost, multiply the number of days by 5 and then add 15'.

a

Pattern 1   Pattern 2   Pattern 3

Pattern number		1	2	3	...
Step 1:					...
Step 2:	No. of matches	6	11	16	...

'To find the number of matches needed, multiply the pattern number by ... and ... ...'.

b

Days hired		1	2	3	4	...
Step 1:						...
Step 2:	Cost (£)	300	350	400	450	...

'To find the cost of tractor hire, multiply the number of days by ... and ... ...'.

c Laying paths

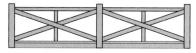

Number of slabs		1	2	3	4	...
Step 1:						...
Step 2:	Path length (m)	2	5	8	11	...

'To find the path length in metres, multiply the number of slabs by ... and ... ...'.

d Making fences

Fence length (m)		1	2	3	4	...
Step 1:						...
Step 2:	Pieces of wood	7	13	19	25	...

e

Minutes used		1	2	3	4	...
Step 1:						...
Step 2:	Total cost (p)	6	15	24	33	...

f Wheels on vehicle

Weight (tonnes)		1	2	3	4	...
Step 1:						...
Step 2:	No. of wheels	4	6	8	10	...

5 Giving the formula

A garden path is made from 2 metre concrete slabs with gaps of 1 metre.
Look at the table.

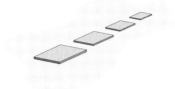

Number of slabs	1	2	3	4	...
×3	3	6	9	12	...
−1: Path length (m)	2	5	8	11	...

The formula for finding the length of the path if you know the number of slabs is:

the length of the path is the number of slabs times 3 then take away 1

or

the length of the path = (number of slabs) × 3 − 1

Exercise 5.1

1 Christmas lights are strung evenly between posts.
The distance between posts is 3 metres.

a Copy and complete the table.

Number of posts	2	3	4	5	6	7		12
	6	9	12					
Distance covered (m)	3	6	9	12				

b Write down a formula for finding the distance covered if you know the number
of posts.

2 Here are some chairs and tables:

1 table seats 6 people.   2 tables seat 10 people.   This is the 3 table arrangement.

a Copy and complete:

Number of tables	1	2	3	4	5	6		11
	4	8						
Number of seats	6	10	14					

b Write down a rule for finding the number of seats if you know the number of
tables.

3 These coins are arranged in patterns.
There is a rule connecting the number of coins and the pattern number.

Pattern 1   Pattern 2   Pattern 3

a Copy and complete:

Pattern number	1	2	3	4	5	6			9
Number of coins	7	12	17						

b Write down a rule for finding the number of coins if you know the pattern number.

4 The sides of these bridges are made from girders.

Girders

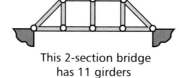
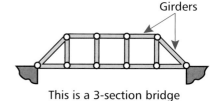

This 1-section bridge has 8 girders   This 2-section bridge has 11 girders   This is a 3-section bridge

a Copy and complete:

Number of sections	1	2	3	4	5	6			9
Number of girders	8	11							

b Write down a rule for finding the number of girders if you know the number of sections.

5 A garden path is made from 3 metre slabs with gaps of 1 metre between them.

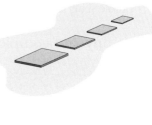

a Copy and complete:

Number of slabs	1	2	3	4	5	6		9
Path length (m)	3	7	11					

b Write down a formula for finding the path length if you know the number of slabs.

6 These designs are made from matches.
There is a rule connecting the number of matches used and the design number.

    ...

Design 1   Design 2   Design 3   Design 4

a Copy and complete:

Design number	1	2	3	4	5	6	12
Number of matches							

b Write down a formula for finding the number of matches if you know the design number.

Challenge

A Starting with Design 1, make your own pattern of designs.

Draw a table showing the number of matches for each of the first six designs.

How many matches are needed for design 20?

Write down a rule for finding the number of matches if you know the design number.

Design 1

B Repeat part **A** but now with your own design 1.

6 Using formulae

Example   Here is a **formula** for finding the cost of hiring a minibus:

$$\text{cost } (£) = (\text{number of days} \times 25) + 100$$

How much does it cost a school to hire the minibus for a full week (7 days)?

Cost $(£) = (7 \times 25) + 100 = 175 + 100 = £275$

It costs £275 to hire the minibus for a week.

Exercise 6.1

1 a The thawing time if you leave a frozen turkey in the fridge is:

thawing time in days = (turkey's weight in pounds) ÷ 4

Find the thawing time for a turkey that weighs:

   i 8 pounds   **ii** 24 pounds   **iii** 18 pounds

b The cooking time for a turkey is given by:

cooking time in hours = (turkey's weight in pounds ÷ 4) + 1

Find the cooking time for a turkey that weighs:

   i 16 pounds   **ii** 28 pounds   **iii** 10 pounds

2 Freddie's Fencing Service calculates the time to build a fence using this formula:

number of minutes = (length of fence in metres) × 15 + 30

a Calculate how long, in hours, it will take to build a fence of length:

   i 2 metres   **ii** 6 metres

b How long, in hours and minutes, will it take to build a fence of length:

   i 4 metres   **ii** 3 metres   **iii** 7 metres?

3 It is very difficult to tell a lobster's age.
This formula is sometimes used:

lobster's age in years = (lobster's weight in pounds) × 4 + 3

How old is a lobster that weighs:

   a 1 pound   **b** 3 pounds   **c** 2 pounds   **d** $\frac{1}{2}$ pound?

4 Cyclists can estimate their time for a 25-mile race from their time for a 10-mile race using:

25-mile time in minutes = (10-mile time in minutes) × 2·5 + 1

Find the 25-mile time, in hours and minutes, for a cyclist whose 10-mile time is:

   a 30 minutes   **b** 24 minutes   **c** 26 minutes

5 Use your calculator and the conversion formulae given to answer these questions.

   a Number of feet = (number of metres) × 3·25
      Change these to feet:
      i 5 metres   **ii** 10 metres   **iii** 35 metres   **iv** 50 metres

   b Number of miles = (number of kilometres) × 0·62
      Change these to miles:
      i 5 km   **ii** 10 km   **iii** 45 km   **iv** 500 km

   c Temperature in °F = (temperature in °C) × 1·8 + 32
      Change these to °F:
      i 15 °C   **ii** 8 °C   **iii** 100 °C   **iv** 0 °C   **v** 27 °C

6 A dog's age can be compared with a 'human' age using the formula:

dog's 'human' age = (dog's age − 2) × 5 + 25

Find the 'human' age for a dog aged:

a 2 years

b 4 years

c 10 years

d 15 years

7 This formula is used to calculate a child's dose of medicine if you know the adult's dose:

child's dose in ml= (adult's dose in ml × child's age in years) ÷ (child's age in years + 12)

Find the child's dose in millilitres when:

a the adult's dose is 10 ml and the child is 3 years old

b the adult's dose is 15 ml and the child's age is 8 years

c the adult's dose is 36 ml and the child's age is 6 years.

Challenge

Solar panels are tilted to catch the sunlight.
They need to be adjusted each month.

The formula for the amount of tilt in January is calculated from the formula:

number of degrees tilt = latitude of place + 15°

Kim lives in Aberdeen.
Her friend Zoe lives in Edinburgh.
Another friend, Mia, lives in London.
The table gives the latitude of each place.

Place	Latitude
Edinburgh	56°
London	51·5°
Aberdeen	57°

Find the number of degrees tilt each person should use for their solar panels.

◀◀ RECAP

Number patterns

You should be able to use a rule to continue a number pattern:

Examples

$\boxed{+\,2}\!\!\rangle$ 8, 10, 12, 14, …

$\boxed{-\,2}\!\!\rangle$ 8, 6, 4, 2, …

$\boxed{\times\,2}\!\!\rangle$ 8, 16, 32, 64, …

$\boxed{\div\,2}\!\!\rangle$ 8, 4, 2, 1, …

Changing number patterns

You should be able to find a rule which changes one number pattern into another.

Examples   **One-stage rule**   1, 2, 3, 4, … becomes 5, 6, 7, 8, …
when the rule is: 'add 4'.

         Two-stage rule   1, 2, 3, 4, … becomes 3, 5, 7, 9, …
when the rule is: 'multiply by 2 and then add 1'.

Formulae

You should be able to find a formula to describe a situation.

Example

Formula:
The number of chairs equals the number of tables times 2 plus 4.

You should be able to use a formula expressed in words.

Example   If you have 25 tables you can seat $25 \times 2 + 4 = 50 + 4 = 54$ people.

1 a Draw the next two missing pictures.   **b** Find the next two missing numbers.

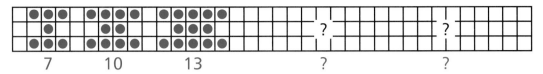

2 For each number pattern:
   i find a rule   **ii** give the two missing numbers.
   a 2, 6, 10, 14, ?, ?   **b** 35, 31, 27, 23, ?, ?   **c** 7, 15, 23, 31, ?, ?

3 Copy and complete the tables:

a

Counting numbers	1	2	3	4
Pattern				

Rule +9

b

Counting numbers	1	2	3	4
Pattern				

Rule ×7

4 Find the missing rule:

a

Counting numbers	1	2	3	4
Pattern	5	6	7	8

Rule ?

b

Counting numbers	1	2	3	4
Pattern	8	16	24	32

Rule ?

5 Find the rule which changes:
   a 1, 2, 3, 4, … into 12, 13, 14, 15, …   **b** 1, 2, 3, 4, … into 4, 8, 12, 16, …
   c 8, 9, 10, 11, … into 1, 2, 3, 4, …   **d** 1, 2, 3, 4, … into 10, 20, 30, 40, …

6 Copy and complete these tables:

a

Counting numbers	1	2	3	4	5
Step 1: ×6					
Step 2: −5					

b

Counting numbers	1	2	3	4	5
Step 1: ×3					
Step 2: +11					

7 Find the two-step rule that changes 1, 2, 3, 4, 5, … into:
   a 2, 5, 8, 11, 14, …   **b** 7, 13, 19, 25, 31, …   **c** 1, 10, 19, 28, 37, …

8 Write down the two-step rule to find the cost if you know the number of
   kilometres:
   a Taxi Hire   **b Luxury Limo Hire**

Kilometres	1	2	3	4
Step 1: ×?				
Step 2: ?: Cost (p)	150	250	350	450

Kilometres	1	2	3	4
Step 1: ×?				
Step 2: ?: Cost (£)	17	26	35	44

REVISE

9 These designs are made from matches.
There is a rule connecting the number of matches used and the design number.

    ...

Design 1   Design 2   Design 3   Design 4

a Copy and complete this table:

Design number	1	2	3	4	5	6	⋯	12
Number of matches								

b Write down a rule for finding the number of matches if you know the design number.

10 The best distance between loudspeakers in a room is given by:

speaker spacing in metres = (ceiling height in metres − 2) × 2

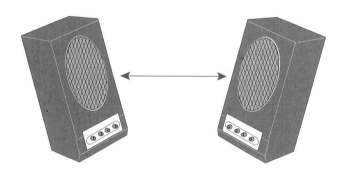

How far apart should you place speakers if the ceiling height is 3·5 metres?

REVISE

8 Percentage in money

We do percentage calculations all the time in everyday life.

It could be when we are shopping, checking our pay or deciding which bank to use.

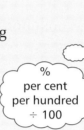

% per cent per hundred ÷ 100

1 Review

◀◀ Exercise 1.1

1 Stevie is doing money calculations. His calculator gives him the answers shown. Write down these amounts of money correctly using the £ sign.

a

25.05

b

240.7

c

7.25

d

8.4

e

12.5

f

27

2 Write these answers in (i) pounds (ii) pence. Each is an amount of money in pounds (£).

a

0.47

b

0.07

c

0.3

3 Calculate:

  a £68 ÷ 2   **b** £39 ÷ 3   **c** £72 ÷ 4   **d** £60 ÷ 5

  e £140 ÷ 10   **f** £73 ÷ 2   **g** £82 ÷ 5   **h** £95 ÷ 10

  i £22 ÷ 4   **j** £124·20 ÷ 3   **k** £15·48 ÷ 2   **l** £7·40 ÷ 4

  m £7·85 ÷ 5   **n** £251·60 ÷ 10

4 Work out:

a $\frac{1}{2}$ of £36   b $\frac{1}{3}$ of £69   c $\frac{1}{5}$ of £750

d $\frac{1}{4}$ of £288   e $\frac{1}{10}$ of £540   f $\frac{1}{2}$ of £84·26

g $\frac{1}{4}$ of £32·60   h $\frac{1}{3}$ of £71·16   i $\frac{1}{5}$ of £37·35

j $\frac{1}{2}$ of £85   k $\frac{1}{4}$ of £75   l $\frac{1}{10}$ of £107

5 Write these percentages as common fractions, e.g. $11\% = \frac{11}{100}$.

a 77%   b 9%   c 13%   d 49%   e 51%   f 67%

6 Write these percentages as common fractions in their simplest form.

a 25%   b $33\frac{1}{3}\%$ c 50%   d 20%   e 10%   f 1%

7 Write these percentages as decimal fractions.

a 42%   b 63%   c 57%

d 9% (careful!)   e 7%   f 70%

---

2 Finding percentages without a calculator

Reminder

$50\% = \frac{1}{2}$	$33\frac{1}{3}\% = \frac{1}{3}$	$25\% = \frac{1}{4}$
$20\% = \frac{1}{5}$	$10\% = \frac{1}{10}$	$1\% = \frac{1}{100}$

Example 1   Find $33\frac{1}{3}\%$ of £45.

$33\frac{1}{3}\%$ of £45 $= \frac{1}{3}$ of £45 $=$ £45 $\div$ 3 $=$ £15

Example 2   Calculate 25% of £350.

25% of £350 $= \frac{1}{4}$ of £350 $=$ £350 $\div$ 4 $=$ £87·50

Exercise 2.1

1 Work out:

a 25% of £40   b $33\frac{1}{3}\%$ of £3000   c 10% of £490

d 20% of £55   e 1% of £600   f 50% of £82

2 Calculate:

a $33\frac{1}{3}\%$ of £72   b 10% of £520   c 50% of £34

d 1% of £7200   e 20% of £45   f 25% of £76

3 Find:

a 10% of £23·30   b 20% of £173·65   c 50% of £620·52

d 25% of £9·04   e 1% of £185   f $33\frac{1}{3}\%$ of £62·25

4 Calculate:

   a 20% of £175·50   **b** 1% of £860   **c** $33\frac{1}{3}$% of £85·20

   d 10% of £76   **e** 50% of £247·40   **f** 25% of £6·80

5 Work out:

   a 1% of £381   **b** 2% of £381   **c** 4% of £381   **d** 8% of £381

6 A car salesman sold cars worth £27 000 in one week.
He has 1% of this total added to his pay.
How much is added to his pay?

7 Charles and Dave were busking in the city centre.
They counted up the money at the end of the day.
They had earned £74.
They decided that they should each get 50% of the £74.

   a How much does each person get?

   b The next day Charles and Dave were joined by Christine.
The three of them earned £216.
They decided that each person should get $33\frac{1}{3}$%.
What is $33\frac{1}{3}$% of £216?

8 George got a job selling programmes for the big game.
His little brother Colin came along to help.
George earned £37·70.
He decided to give Colin 20%.
How much did he give Colin?

9 Sammy earned £248·50 a week on her window cleaning round.
She lost 10% of her round to a new window cleaner called Bill.
How much of her weekly earnings did Sammy lose?

Exercise 2.2

1 A menswear shop is having a sale.

   a How much do you save when you buy:

      i   a tie
      ii   a pair of trousers
      iii   a shirt
      iv   a blazer?

   b How much do you pay for:

      i   a tie
      ii   a pair of trousers
      iii   a shirt
      iv   a blazer?

SALE
10% off these prices
Trousers £35
Shirts £28
Blazers £42·50
Ties £20

2 The value of Robert and Janet's house increased by 25% in one year. The house was worth £76 000 at the beginning of the year.

   a By how much did the house increase in value?

   b How much was the house worth at the end of the year?

3 Steven was usually paid £38·50 for working on a Saturday morning. On the Saturday morning of Festival Week he was paid 50% extra.

   a How much extra was he paid?

   b What was his total pay for the Saturday morning of Festival Week?

4 Monty spent £73·50 at a garden centre. Using his 'Gardeners' Friend' discount card he was able to get a discount of $33\frac{1}{3}\%$.

   a How much was the discount?

   b How much did he have to pay?

Challenge

   a Can you think of a way to work out 5% by first working out 10%?

   b Would this help you work out 15%?

   c Could you work out 30% by first working out (i) 10% (ii) 15%?

   d Think of two ways you could work out 40%.

---

3 Finding percentages with a calculator

Example   Calculate 7% of £500.

        7% of £500 = $\frac{7}{100}$ of £500

        500 ÷ 100 gives 1%.

        Multiply this by 7 to get 7%.

        We would key in:

        (500) (÷) (100) (×) (7) (=) | 35 |

        7% of £500 = £35

Exercise 3.1

1 Work out:

    a 9% of £300   **b** 4% of £75   **c** 5% of £260

    d 24% of £150   **e** 17% of £600   **f** 12% of £4800

    g 8% of £225   **h** 15% of £180   **i** 80% of £95

    j 32% of £50   **k** 89% of £800   **l** 41% of £1000

2 Ray Kittin is a footballers' agent.
One of his star players earns £5000 a week.
Ray gets 4% of the £5000.
How much does he get?

3 A market stall sells Christmas trees.
On one day 71% of the money spent was for Norway Spruce trees.
A total of £3500 was spent on that day.
How much was spent on Norway Spruce?

4 £250 was spent on a Christmas dance.
18% of the £250 was for decorations.
How much did the decorations cost?

5 Work out:

    a 7% of £45   **b** 3% of £124   **c** 6% of £83

    d 13% of £785   **e** 62% of £374   **f** 35% of £567

    g 37% of £86   **h** 53% of £7   **i** 9% of £65

    j 82% of £530 (Careful)   **k** 17% of £280   **l** 43% of £620

    m 45% of £752   **n** 3% of £70   **o** 11% of £950

6 A gardener spends £51 on plants for his border.
17% of this money is spent on a special plant.
How much did he spend on this plant?

7 Peter sells magazines in a city centre.
He gets to keep 85% of the money.
One day he sells magazines worth £134.
How much does he keep?

Exercise 3.2

1 Work out and round to the nearest penny:
   a 14% of 80p   **b** 27% of 70p   **c** 55% of 42p
   d 6% of 48p   **e** 35% of 83p   **f** 37% of 25p

2 A big supermarket charges 35p for a tin of baked beans.
   The price is to rise by 7%.
   a What is 7% of 35p rounded to the nearest penny?
   b What is the new price of a tin of beans?

3 A chocolate bar costs 42p. The price is to rise by 7%.
   a What is 7% of 42p rounded to the nearest penny?
   b What is the new price of the chocolate bar?

4 Bread costs 63p a loaf. Some loaves are getting near their
   sell-by date. The manager decides to reduce the price by 80%.
   a What is 80% of 63p?
   b What is the reduced price?

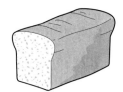

5 Work out and round to the nearest penny:
   a 13% of £25·60   **b** 32% of £65·20   **c** 83% of £925·40
   d 5% of £72·25   **e** 28% of £420·70   **f** 7% of £130·65

6 Davina charges £23·50 per hour to hire her karaoke.
   She decides to increase this charge by 5%.
   a How much is the increase (rounded to the nearest
      penny)?
   b How much will Davina charge after the increase?

7 Harry charges £18·75 per hour to hire his disco.
   He gives a 15% discount to friends and family.
   A friend hires the disco for three hours.
   a How much would it cost if there was no discount?
   b How much discount would there be (rounded to the nearest penny)?
   c How much will Harry's friend pay to hire the disco?

8 Patricia is fixing her living-room floor.
   It costs her £23·25 per hour to hire a sander.
   a How much will it cost to hire the sander for 5 hours?
   b A week later the prices go up by 8%.
      How much will it cost to hire the sander for 5 hours now?

Challenge

Tam is buying a new computer.
It costs £799.
VAT is charged at 17·5%.

a Calculate: **i**   10% of 799
                     ii 5% of 799 (by halving 10%)
                     iii 2·5% of 799 (by halving 5%)

b Use your three answers to work out the VAT on Tam's computer.

c How much in total must Tam pay?

---

4 Simple interest

When we keep money in a bank, it
usually earns us interest.
The interest is a percentage of the
amount of money we have in the bank.
If we are saving money we usually put it into a
deposit account or a **savings account**.

Example   Lynsey has savings of £345 in a deposit account.
              The interest rate is 7% p.a.   (Remember: p.a. means per year)
              How much interest will she receive for one year?

              Interest = 7% of £345 = 345 ÷ 100 × 7 = £24·15

Exercise 4.1

1 The Friendly Bank offers an interest rate of 4% p.a.
   on its deposit account.
   Work out how much interest you would receive in a
   year for:

   a £100   **b** £400   **c** £50

   d £300   **e** £725   **f** £3000

> **THE FRIENDLY BANK**
>
> Deposit account
> interest rate
> **4% p.a.**

2 Julie is going to put some money into a bank.
   She thinks that the Bank of Goodwill is the best.
   She will be paid interest after one year.

   How much interest will she get if she puts in:

   a £100   **b** £40   **c** £79

   d £53   **e** £125   **f** £670?

> **Bank Of Goodwill**
>
> Savings Account
> interest rate
> **6% p.a.**

3 The Southern Crag Bank has an interest rate of 7% p.a.
In a year how much interest would you receive for:

    a £100   **b** £700   **c** £325   **d** £75   **e** £120?

4 Tracy has £783 to bank.
How much interest will she get if the interest rate is:

    a 3% p.a.   **b** 5% p.a.   **c** 6% p.a.   **d** 4% p.a.   **e** 7% p.a.?

5 The Young Enterprise Savings Bank gives an interest rate of 10%.
How much interest will the bank pay if the following amounts are invested for a year?

    a £100   **b** £620   **c** £5000

    d £730   **e** £337   **f** £7488

    g £3·50   **h** £421·70   **i** £60·70

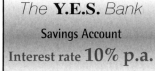

The **Y.E.S.** Bank

Savings Account

Interest rate **10% p.a.**

Challenge

The bank's accountants warn that the bank's interest rate is too generous.

The bank is advised to cut its interest rate to 5%.

Use your answers to question **5** to calculate the amount of interest that will be paid at the new rate.

Exercise 4.2

1 Kayleigh, Abbie and Callum are given £500 each by their Auntie Angela.
They put their money in different banks.
Calculate their interest after one year if:

    a Callum uses the Tweedsdale

    b Kayleigh uses the Speyside Vault

    c Abbie chooses the Tay Savings Bank.

Bank	Interest rate
Tweedsdale Bank	5%
Speyside Vault Bank	4%
Tay Savings Bank	7%

2 Sylvia has £529·68 in her savings account.
The interest rate is 6% p.a.

    a How much interest will she receive in a year ? (Round it to the nearest penny.)

    b How much will she have in her account at the end of the year?

3 Glen Tennis Club has £1365·52 in its savings account.
The interest rate is 4%.

    a How much interest would the club receive after one year?
(Round to the nearest penny.)

    b How much would the club have in its savings account after one year?

4 The Glenbarton Christmas Lights Committee has a savings account at the bank.
They take out the interest every year and use it for the Christmas lights.
There is £3257·72 in the account. The interest rate is 5% p.a.
   a How much will the interest be (rounded to the nearest penny)?
   b They have also gathered £279·38 in collections.
   Add this to the interest to find out how much they can spend on the lights
   this year.

People can also borrow money from the bank.
When this happens interest has to be paid *to* the bank.
These interest rates are usually higher than those for a savings account.

5 Matthew has taken out a loan to pay for his new mountain bike.
The bike cost £530 and the interest rate is 8% p.a.
How much is the interest for one year?

6 A football club owes the bank £9000.
The interest rate is 6%.
How much interest does the football club pay in a year?

5 Wage increases

Exercise 5.1

1 Georgina works in a craft shop.
She is paid £4·50 an hour.
On her next birthday she will be due a pay increase of 6%.
   a By how much will her pay increase?
   b How much will she be paid per hour?

2 Daniel has a summer holiday job.
He is a guide at a stately home.
At first he is paid £5·20 an hour.
After 3 weeks his pay is increased by 10%.
   a How much is this pay increase?
   b How much will he now be paid per hour?

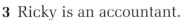

3 Ricky is an accountant.
He earns £38 000 a year.
Next year his pay will go up by 3%.
   a By how much will his pay go up?
   b What is his new annual salary (yearly pay)?

4 Grace works on a farm.
She is paid £4·80 an hour.
When she works at the weekend she is given 33⅓% extra.
   a How much extra per hour does she get at weekends?
   b What is her hourly rate at weekends?

5 Karen is a trainee hair stylist.
She is paid £150 per week.
On her seventeenth birthday her pay goes up by 7%.
a How much is the pay increase?
b What is her new weekly pay?

6 Julien is a footballer.
His weekly wage is £290.
His club has money problems.
They are asking him to take a pay cut of 8% for next season.
a How much is the pay cut?
b North County are offering to pay him £270 per week.
Would he better off if he transferred to North County?

◀◀ RECAP

Percentages
You should be able to work out common percentages of money without using a calculator.
Change them to simple fractions first.

$1\% = \frac{1}{100}$ (for 2% double 1%, for 4% double 2%, for 8% double 4%)
$10\% = \frac{1}{10}$ (for 5% halve 10%)
$20\% = \frac{1}{5}$
$25\% = \frac{1}{4}$
$33\frac{1}{3}\% = \frac{1}{3}$
$50\% = \frac{1}{2}$

You should be able to work out whole number percentages of money using a calculator.

Example   Calculate 27% of £35.
            27% of £35 = 35 ÷ 100 × 27 = £9·45

Interest
You should be able to calculate how much interest is earned over one year.

Example   The interest rate is 6% p.a.   (p.a. means per year)
            How much interest would you receive over a year for £45?
            6% of £45 = 45 ÷ 100 × 6 = £2·70

Wage increases (and decreases)
You should be able to calculate a wage rise.

Example   Jean is paid £270 per week.
            She is given a pay rise of 3%.
            How much is her pay rise?
            3% of £270 = 270 ÷ 100 × 3 = £8·10

1 Work out:
  a 50% of £23·30   **b** 10% of £173·60   **c** 25% of £620·52
  d 20% of £9·25   **e** 1% of £185   **f** $33\frac{1}{3}$% of £72·51

2 Find:
  a 1% of £725   **b** 2% of £725   **c** 4% of £725   **d** 8% of £725

3 Calculate:
  a 10% of £650   **b** 5% of £650   **c** 10% of £327·60
  d 5% of £327·60   **e** 10% of £4238 **f** 5% of £4238

4 Karen is a singer. She gives 5% of her
fee to her agent and keeps the rest.
One night she is paid £175 for a show.

  a How much does she give to her agent?

  b How much does she keep for herself?

5 Calculate these, rounding to the nearest penny when you need to.
  a 6% of £100   **b** 17% of £88   **c** 53% of £620
  d 34% of £646   **e** 77% of £720   **f** 98% of £870
  g 8% of £358·42   **h** 28% of £36·70 **i** 74% of £375·84

6 James sells £758 worth of clothes in one day.
27% of this money came from shirts he sold.
How much money was spent on shirts?

7 Paul started work at 16. His pay was £170 a week.
On his seventeenth birthday he received an 8% increase.

  a How much was the increase?

  b What was his weekly pay after his seventeenth birthday?

8 Marie puts £325 into her bank. The interest rate is 4% p.a.

  a How much interest will Marie receive after a year?

  b How much will she have in her account then?

9 Tiling and symmetry

Many objects are designed to have symmetry.
Things often look more attractive if they are symmetrical.
Would you buy a lopsided car?

Sometimes symmetry appears in nature.
It can make a piece of scenery even more beautiful.

1 Review

◀◀ Exercise 1.1

1 Which of these pictures has a line of symmetry?

a    b   c    d    e

2 Sketch each shape and mark in its line(s) of symmetry.

Square

Isosceles
triangle

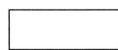

Rectangle

3 Copy the scale on this thermometer.

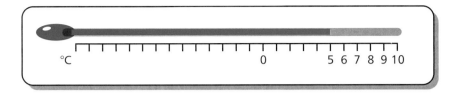

Complete the scale by filling in the missing numbers.

2 Co-ordinates

Using coordinates is a good way to describe position.

Here is a **Cartesian** diagram.
This diagram has an *x* axis and a *y* axis.
The *x* axis and the *y* axis meet at the point (0, 0).
This point is called the **origin**. 'Origin' means 'beginning'.

In this diagram the point A has coordinates (6, 2).
This means that from the origin (the beginning) you have to go 6 to the right and 2 up to get to A.

The point B has coordinates (3, 5).
We say that the *x* coordinate of B is 3 and the *y* coordinate of B is 5.

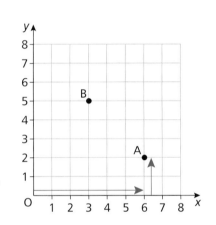

Exercise 2.1

1 Look at this Cartesian diagram.
  a Copy and complete:
    The coordinates of K are (…, …)
  b Write down the coordinates of:
    i A   **ii** B
    iii C   **iv** D
    v E   **vi** F
  c Name the points at:
    i (6, 4)   **ii** (7, 7)
    iii (0, 6)   **iv** (3, 7)
  d Name a point with two coordinates the same.
  e Name a point which is on the *x* axis.
  f Name a point which is on the *y* axis.

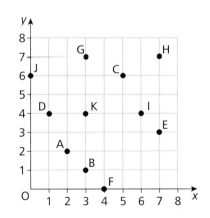

2 a Copy and complete:
The *x* coordinate of A is ...
The *y* coordinate of A is ...

b What is the *x* coordinate of B?

c What is the *y* coordinate of C?

d Name three points which have the same *x* coordinate.

e Name three points with the same *y* coordinate.

f Which point has the largest *x* coordinate?

g Which point has the largest *y* coordinate?

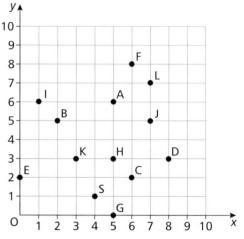

3 Follow these instructions to draw a Cartesian diagram.

a Draw an *x* axis numbered 0 to 8.

b Draw a *y* axis going up from 0 to 8.

c Now plot these points, joining them up as you go.
You should see 'a sign of the times'.
(6, 1), (4, 3), (6, 5), (5, 6), (3, 4), (1, 6), (0, 5), (2, 3), (0, 1), (1, 0), (3, 2), (5, 0), (6, 1)

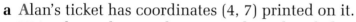

4 At a local agricultural show a paddock is marked out like a Cartesian diagram.
People buy tickets which have coordinates printed on them.
A cow is released into the paddock.
The judge watches carefully to see where the first 'cow-pat' lands.
The person who has the coordinates closest to this spot wins a colour TV.

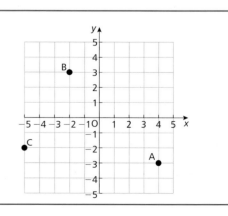

a Alan's ticket has coordinates (4, 7) printed on it.
Write down the coordinates on the tickets belonging to:
i Barbara   **ii** Colin   **iii** Evelyn   **iv** Gavin
v Jimmy   **vi** Stuart   **vii** Zoe

b The first cow-pat lands at (6, 5). Who is the winner?

We can use negative numbers to extend our Cartesian diagram. In the example shown here:

    A has coordinates (4, −3)
    B has coordinates (−2, 3)
    C has coordinates (−5, −2).

A negative *x* coordinate tells us to go left instead of right.

A negative *y* coordinate tells us to go down instead of up.

Exercise 2.2

1 a Write down the coordinates of A.
b Write down the coordinates of B.
c Which point is at (−4, −2)?
d Which point is at (4, 1)?
e Which point is at (−2, −5)?

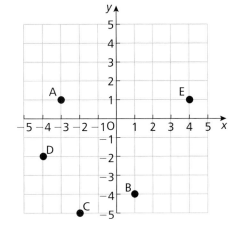

2 The coordinates (−2, 4), (−3, 2) and (−4, −3) spell AYR.
What is spelled by these coordinates?
a (1, 1), (3, 3), (−4, −3), (1, −3), (−5, 3)
b (5, −4), (3, 3), (−2, −1), (−4, 0), (0, 0)
(Remember (0, 0) is always O.)

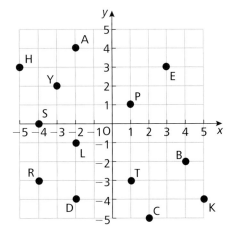

3 Line symmetry

This butterfly has line symmetry.

The left-hand side is a **reflection** of the right-hand side. We say that the left-hand side is the **image** of the right-hand side.

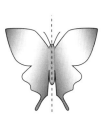

Put a mirror on the dotted line.
You should see exactly the same when the mirror is there as you did before.

The dotted line is called an **axis of symmetry**.

An axis of symmetry splits a picture into two halves.

Each half is the mirror image of the other half.

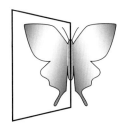

Exercise 3.1

1 One of these drawings does not have line symmetry. Which one?

a

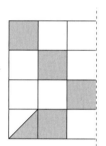

b

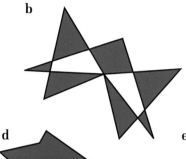

c

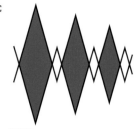

d

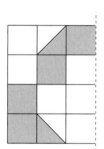

e

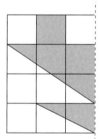

2 Copy each of these diagrams on squared paper.
Then complete each one so that the dotted line is an axis of symmetry.

a

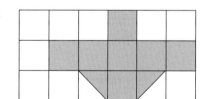

b

c

3 Copy and complete each diagram so that the dotted line is an axis of symmetry.

a

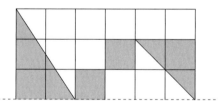

b

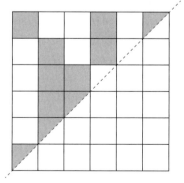

4 Copy and complete each diagram so that the diagonal line is an axis of symmetry.

a

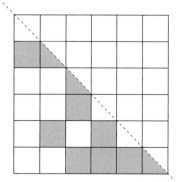

b

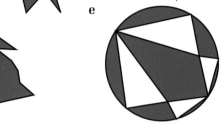

5 Copy and complete each diagram so that each line is an axis of symmetry.
Note that some parts on both sides of the line have been shaded.

a 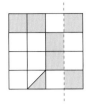 b 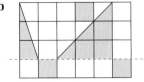 c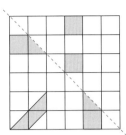

6 Copy theses and draw lines so that the dotted lines are axes of symmetry.

a 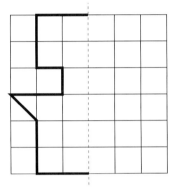 b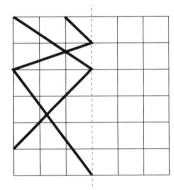

Some shapes have more than one axis of symmetry.

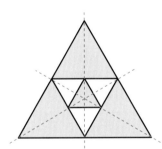 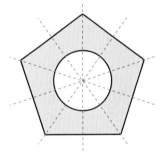

This design has three axes
of symmetry.

This design has five axes
of symmetry.

Exercise 3.2

1 How many axes of symmetry do these designs have?

a b c d 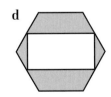 e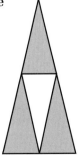

2 First reflect in axis 1, then reflect in axis 2 to create patterns with two axes of symmetry.

a

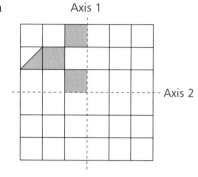

b

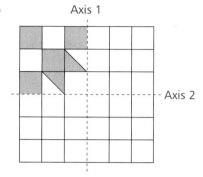

c

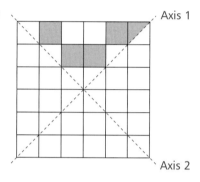

d

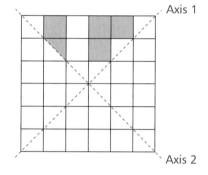

3 Trace these shapes and mark in their axes of symmetry. (It may help you to fold them.)

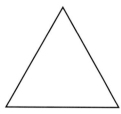

Equilateral triangle
(3 sides the same
length)

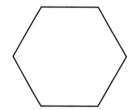

Regular
hexagon

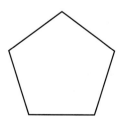

Regular
pentagon

Isosceles triangle
(2 sides the same
length)

4 a Copy this diagram into your jotter.

  b Join these points in order:
    (4, 8), (2, 6), (3, 6), (1, 4), (2, 4),
    (0, 2), (3, 2), (3, 0), (4,0).

  c Complete the diagram so that the dotted line
    is an axis of symmetry.

  d Make a list of the points that you have
    joined in **c**.

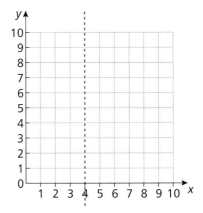

4 Tiling

Reminder:

> A tiling is when identical shapes are arranged to fit together to
> cover a flat surface. They must leave no gaps and have no
> overlaps.
>
> The mathematical name for this is a **tessellation**.

These examples show hexagons tiling in a honeycomb, squares tiling on a
chessboard and rectangles in a wall, but there are many more.

Many shapes will tile with a bit of planning.

When asked to continue a tiling, you should do so until one tile is completely
surrounded.

Exercise 4.1

1 Copy and continue each tiling on squared paper.
Fill the whole area, leaving no gaps and having no overlaps.

a 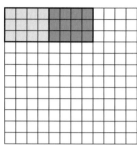   **b** 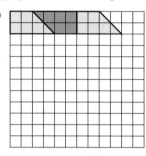   **c**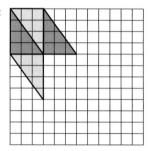

2 Copy and continue these tilings. You should add at least five more tiles.

a 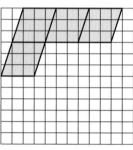   **b**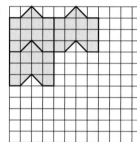

c 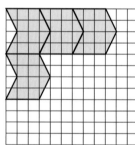   **d**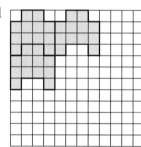

3 Copy and continue these tilings. Note that you have to turn the tiles round.

a 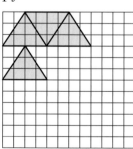   **b**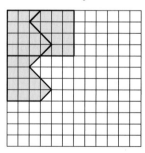

c 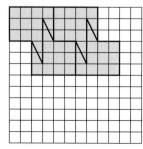   **d**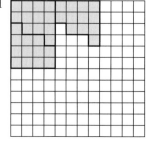

4 Mr Broughton wants to have block paving
laid in his driveway.
He has the choice of four patterns.
Copy and continue each pattern.

a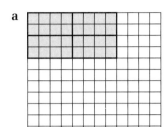

b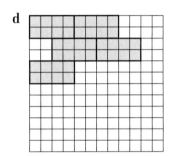

c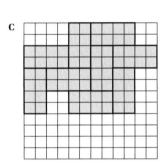

d

5 Mrs Broughton would prefer the driveway done with
hexagonal paving slabs.

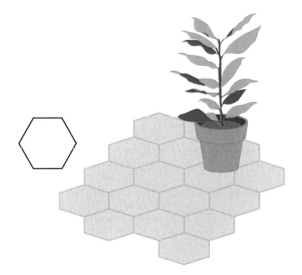

Trace this hexagon. Show how it can fit together to
form a tiling.
Comment on the edges of your tiling.

◀◀ RECAP

Position

A Cartesian diagram can be used to describe position.

The point O, (0, 0) is called the **origin**.

The point A has coordinates (6, 4) because from the origin you go 6 along and 4 up to get to A.

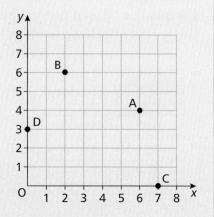

Mirror or line symmetry

This dragonfly has line symmetry.

One side is a reflection of the other side.

When you put a mirror on the axis of symmetry (the dotted line) you see exactly the same as you did before.

You should be able to complete a drawing to give it line symmetry.

Tiling

When identical shapes cover a flat surface without leaving gaps or overlapping we say that the shapes tile.

The mathematical name for this is **tessellation**.

You should be able to continue a tiling.

1 a Write down the coordinates of all the points marked on the diagram.

b On a similar diagram plot the points:
P(4, 7), Q(5, 2), R(2, 2), S(7, 1).

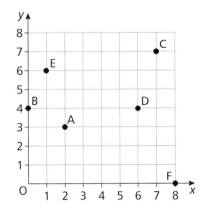

2 Bluedale Rovers are choosing a new strip.
They have decided to have a shirt that is **not symmetrical**.
Which one of these is not symmetrical?

a   **b**   **c**

3 Copy and complete these diagrams so that the dotted line is an axis of symmetry.

a   **b**

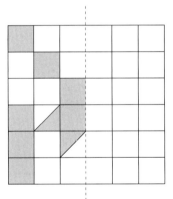 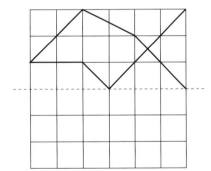

5 Add at least five tiles to each of these tilings.

a   **b**

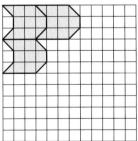 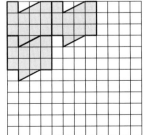

10 Statistics

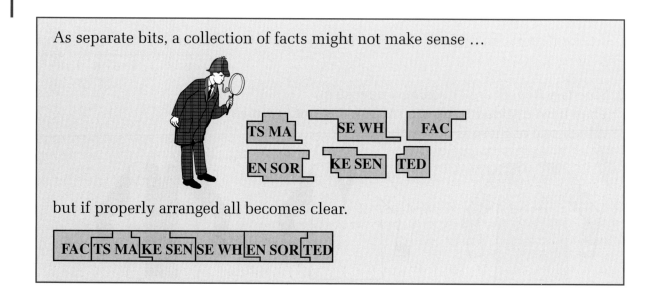

As separate bits, a collection of facts might not make sense ...

TS MA   SE WH   FAC

EN SOR   KE SEN   TED

but if properly arranged all becomes clear.

FAC TS MA KE SEN SE WH EN SOR TED

1 Review

◄◄ Exercise 1.1

1 Alasdair went round his garden counting the snails he found.

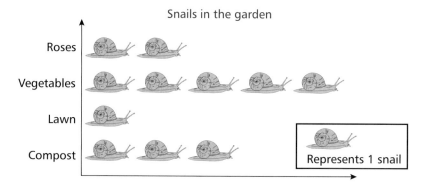

Snails in the garden

Roses

Vegetables

Lawn

Compost

Represents 1 snail

   a Which part of the garden had the most snails?

   b How many snails were in the compost?

   c What kind of diagram is this?

   d What is important about the small box on the right of the diagram?

   e How many snails in total were found?

2 Sandy counted the number of passengers getting off buses arriving at the terminus:

4, 6, 7, 7, 5, 8, 5, 7, 9, 7

a How many passengers were counted?

b How many buses arrived while Sandy was counting?

c Calculate the mean number of passengers per bus.

d What is the mode for these findings?

3 Katrona counted the stamps in her collection.

Place	Europe	UK	USA	Asia	Africa	Others
Number of stamps	50	120	24	12	14	32

She made a chart of the data, but didn't finish labelling it.

a Where were most of her stamps from?

b The label 'Places' should be put on the horizontal axis.
What 'place' name should replace
i A   **ii** B   **iii** C
iv D   **v** E   **vi** F?

c What label should be on the vertical axis?

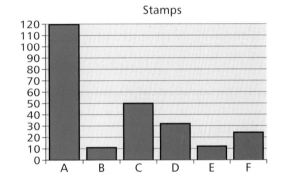

4 Pat and Terri have gone to Ireland for a break.
They make a ready-reckoner to help them convert their pounds (£) to euro (€).

a How many euro would they get for £70?

b Estimate how many euro they would get for £20.

c Terri buys a book for 20 euro.
What is that in pounds (roughly)?

d Pat buys a new bag for 50 euro.
What is this in pounds?

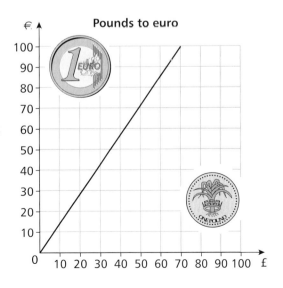

5 Copy each scale and then label the marked divisions.

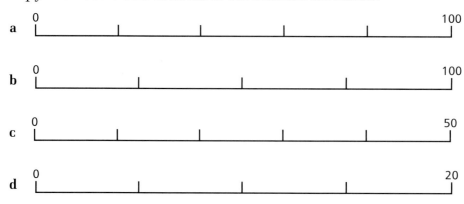

a   0 ⊢——————————————————————————⊣ 100

b   0 ⊢——————————————————————————⊣ 100

c   0 ⊢——————————————————————————⊣ 50

d   0 ⊢——————————————————————————⊣ 20

2 Reading pictograms

Each icon in a pictogram may represent more than one object.

Example   If an anchor, ⚓, is used to represent 5 boats in the harbour, then

we could represent 4 by ⚓ , 3 by ⚓ , 2 by ⚓ and 1 by ⚓

Exercise 2.1

1 Karen counts the boats moored at the Duck Bay Marina each day.

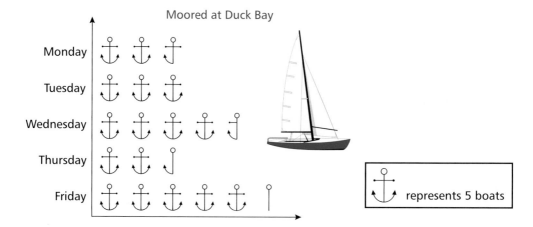

Moored at Duck Bay

⚓ represents 5 boats

a What was the busiest day?

b How many boats were moored on
  i Monday
  ii Thursday?

c How many more were moored on Wednesday than Tuesday?

2 Vicky charts the sale of trees in the weeks before Christmas.
In the diagram, each icon represents ten sales.
So each branch represents one sale.

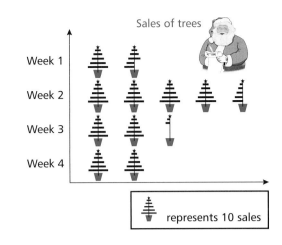

Sales of trees

Week 1

Week 2

Week 3

Week 4

a Which week had the most sales?

b How many sales was this?

c What were the total sales over the 4 weeks?

d How many more trees were sold in week 3 than in week 1?

represents 10 sales

3 Bruce checked the cost of travelling to school from different places.

Getting to school

Langshaw

Heriot

Tweedside

Galabank

Hamfield

a How much does it cost to get to school from
  i Langshaw
  ii Galabank?

b Which is the cheapest journey?

c How much more does it cost to travel from Heriot than from Hamfield?

represents 50p

4 Deirdre checked the number of cars parked in the street in front of her house each hour.

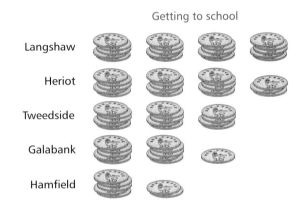

Cars parked in street

8 am

9 am

10 am

11 am

12 noon

a How many were parked at
  i 8 am   **ii** 9 am?

b How many more were parked at 9 am than 10 am?

c i What was the busiest period?
  ii How many cars were parked then?

d If it costs 20p to park for 1 hour, how much money should be collected from 8 am to 12 noon?

represents 2 cars

3 Making pictograms

Exercise 3.1

1 Dave ran a concert for a week.
He sold CDs of the acts at the intervals.
The table shows the sales.

Day	Mon	Tue	Wed	Thu	Fri
Sales	5	10	14	11	18

He wants to use the icon ☆ to represent 5 sales.

a How many icons are needed to represent Tuesday's sales?

b Draw how Wednesday's sales will be represented.

c Draw the pictogram to show the week's sales.

2 Lisa counted the attendance at the concerts.

Day	Mon	Tue	Wed	Thur	Fri
Attendance	50	60	46	32	58

She wants to use the icon ⤬ to represent 10 people.

(Each 'bit' represents 2 people.)

Draw the pictogram to show the week's attendance.

3 Mr Turpie sells double glazing.
He notes the sales of the different types.

Type	Aluminium	Pine	PVC	Others
Sales	20	16	10	9

He uses the icon ⊞ in a pictogram to
represent 4 sales.
Complete his pictogram for him. Remember the key.

Double-glazing sales

Aluminium ⊞ ⊞ ⊞ ⊞ ⊞
Pine
PVC
Others

4 Catriona collects information on the number of coaches that turn up at the big
match each Saturday over 4 weeks.

Week	1	2	3	4
Coaches	20	12	9	13

She decides to use 🚌 to represent 5 coaches.

(So, for example, 🚌 represents 3 coaches.)

Make a pictogram to show her data.

4 Reading bar charts

Remember to check how the scales are numbered.

Exercise 4.1

1 Alasdair McDonald has a farm.
He has to keep a close watch on the weather.
The bar chart shows the number of hours of
sunshine at the farm over 5 weeks.

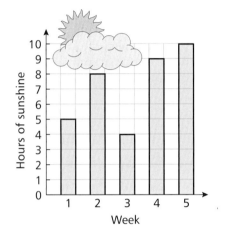

a What was the cloudiest week?

b Which week had most sun?

c What was the total number of hours of sunshine
over the 5 weeks?

d What is the mean number of hours of sunshine
per week?

e Which weeks had less than the mean number
of hours of sunshine?

2 Big Al and Chunk are bird spotters.
Their records for one day have been recorded on the bar chart below.

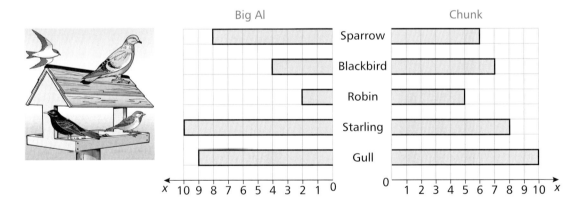

a Who saw the most
  i sparrows
  ii gulls?

b How many more robins did Chunk see?

c How many starlings were spotted in total?

d Which bird was spotted the least in total?

e Who saw the most birds?

3 Mr Blake studies the absentee figures for the 2 weeks before the school holidays.

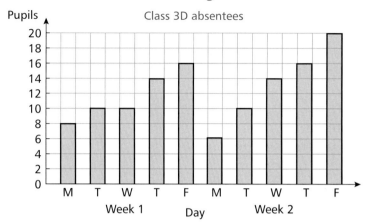

a On which day did the number absent first go over 10?

b How many more were absent on the second Wednesday than the first?

c What was the total number of pupils absent in:
   i the first week
   ii the second week?

d There are 30 pupils on the roll in class 3D.

   How many were *present* on the last Friday?

4 Jen is a painter and decorator.
She records how many tins of each colour of paint she uses over a couple of weeks.

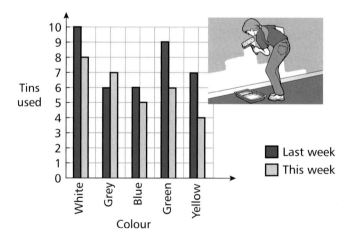

a How many tins of white did she use last week?

b How many tins of blue did she use this week?

c How many more tins of grey did she use this week than last?

d Which week was she busier? Give a reason.

e Which colour did she use the most over the 2 weeks?

5 Making bar charts

Exercise 5.1

1 Dorothy does a survey to find her customers' favourite crisp flavours.

Flavour	Plain	Tomato	Cheese	Bacon	Vinegar
Votes	45	25	30	35	10

  a Copy and complete the bar graph to get a picture of the data.

  b What was the modal flavour (the mode)?

  c How many votes were cast?

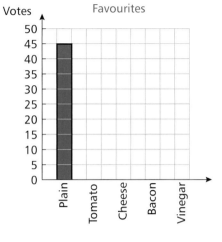

2 Eddie noted the length of the commercial breaks on a television station over several days.

Time (min)	2	3	4	5	6
Number of breaks	20	18	12	8	6

  a Copy and complete the bar graph.

  b What was the modal time?

  c How many breaks were timed?

  d How many breaks lasted more than 3 minutes?

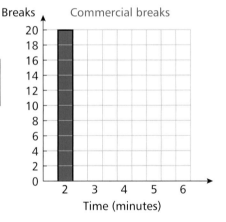

3 Lesley wanted to compare last week's rainfall with this week's.
The table gives this week's figures.

Day	Mon	Tue	Wed	Thu	Fri
Rainfall (mm)	3	7	4	4	9

  a Copy and complete the bar chart.
Last week's figures are already drawn.

  b Which week was wetter?

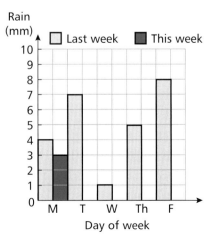

6 Trends

A graph can help us spot **trends**.
We can see how one thing changes as another changes, usually
how one thing changes with time.

Example 1   The graph shows how the value
of Bryan's house has changed
over the years.
Describe the trend.

As time passes we can see that
the trend of the house's value is
to increase.

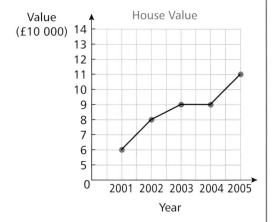

Example 2   A cup of tea sits on a table.
Its temperature over a spell of
time is graphed.
Comment on the trend.

As time passes the temperature
drops.
The cup of tea is cooling down.

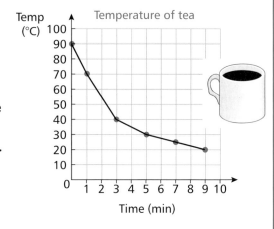

Exercise 6.1

1 Kirsty and Craig build a snowman.
The graph shows the height of each snowball
as they roll it.

a How high is a snowball that has been rolled
   i   2 metres   **ii**   6 metres?

b How far would they need to roll a snowball
to make it 100 cm tall?

c Describe the trend by copying and
completing the sentence:
'The further you roll a snowball the … it gets.'

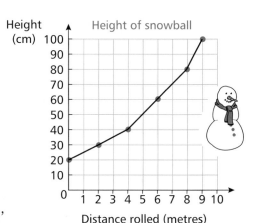

2 Laura is doing an experiment.
As an ice cube melts, she dries and
weighs the ice left.
The graph shows her results.

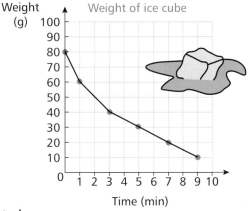

Weight of ice cube

a What was the weight of the cube at the
start?

b What was its weight after 3 minutes?

c When did it weigh 20 g?

d Describe the trend by copying and
completing:
'The longer you wait, the … the ice cube gets.'

3 Alfie is in hospital for a check-up.
The nurse keeps a note of his
temperature on a chart.

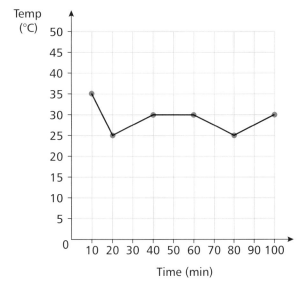

a What was Alfie's temperature after
i 20 minutes
ii 60 minutes
iii 80 minutes?

b Describe the trend by copying and
completing:
'Alfie's temperature is fairly steady,
sitting mostly around …'

7 Seeing the trend in a table

Exercise 7.1

1 The Last-Bus Express only drops people off as it travels away from the city centre.
The table shows the number of passengers on the bus at each fare stage.

Stage	1	2	3	4	5	6
Passengers	46	34	22	18	12	2

a How many people were on the bus at stage 1?

b How many were on at stage 2?

c How many got off between stage 1 and stage 2?

d Copy and complete the sentence to describe the trend:
'As the stage number gets bigger, the number of passengers gets …'

2 It was a very hot day. The table shows the temperature (°F) in the shade over several hours.

Temperature through the afternoon

Time of day	12 noon	1 pm	2 pm	3 pm	4 pm	5 pm
Temperature (°F)	101	95	93	84	82	80

a What was the temperature at
  i 1 pm
  ii 4 pm?

b Was this a rise or fall in temperature?

c Between what times was the temperature likely to be 90 °F?

d Copy and complete the sentence to describe the trend:
  'The later in the afternoon it gets, the … it gets.'

3 The heights of little seedlings are measured for several days as they germinate.

Height of seedlings

Day	1	2	3	4	5	6	7	8
Height (mm)	2	4	5	7	14	18	22	32

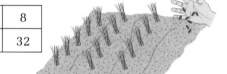

a What was the height on
  i day 3
  ii day 6?

b Was this an increase or decrease in height?

c Between what days was the height likely to be 20 mm?

d Copy and complete the sentence to describe the trend:
  'The longer the seedling has been growing, the … it gets.'

4 Tom hit the golf ball.
Using a video camera, Toni recorded the height of the ball in flight.

Height of ball

Distance from tee (m)	10	20	30	40	50	60	70	80	90
Height of ball (m)	20	38	46	51	48	40	26	11	0

a What height was the ball when it was 30 m from the tee?

b How far from the tee was the ball when its height was recorded as 38 m?

c When else would it have been at this height? (Hint: coming down)

d Copy and complete:
  'As it moved away from the tee the ball rose then fell,
  reaching a highest point of around … m.'

8 Using scatter graphs for trends

Scatter graphs can be used to highlight possible trends.
This scatter graph shows the sales of hot cross buns in a shop as Easter approached.

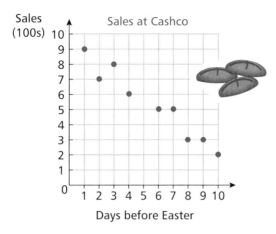

Exercise 8.1

1 Use the scatter graph above to answer these questions.

  a How many sales were recorded **i** 10 days **ii** 4 days before Easter?

  b Is the number of sales on the increase or decrease as Easter approaches?

  c This is called a seasonal trend because sales depend on the season, or time of year.
  When are the sales of

  i gloves
  ii ice creams
  iii umbrellas
  iv parasols

  likely to increase?

2 Several students were asked how long they had studied for a test.
A scatter graph was made of their replies against their score.

  a One person studied for 6 hours.
  What did she score?

  b Two people said they studied for 7 hours.
  What did each score?

  c Copy and complete: 'In general the trend is the more you study, the … you score.'

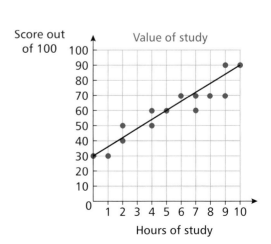

3 George does a survey on how long it takes each person in his office to get to work in the morning. He makes a scatter graph of the time taken against the distance the person lives from the office.

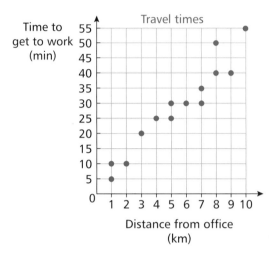

a How long does it take to get to work when you are
  i 2 km   **ii** 9 km away?

b Two people lived 7 km away.
  How long did it take each of them to get to work?

c There is a connection between the distance from work and the time it takes to get there. Describe it.

4 Anne studied the movements of taxis in town.
  She picked various places around the town and counted the number of taxis which passed her in 10 minutes.

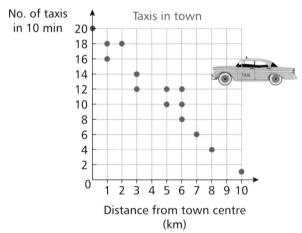

a How many taxis did she see in the town centre? (Hint: distance = 0 km)

b How many taxis did she see 7 km from the centre?

c In general, what is the trend?
  (What happens to the number of taxis she counts
  in 10 minutes as she moves away from the centre?)

9 Using frequency tables

To compare data you need to sort it first.

Example   Gordon compares the performance of a golf ball and a tennis ball.
Both are dropped from the height of a metre.
The size of the bounce is measured.
This is repeated 50 times.

Golf ball

Height of bounce (cm)	50	51	52	53	54	55
Number of drops	4	9	21	12	3	1

Tennis ball

Height of bounce (cm)	46	47	48	49	50	51
Number of drops	1	5	16	22	4	2

Now they can be compared.

Exercise 9.1

1 Answer the following using the frequency tables above.

  a What was the lowest recorded bounce of
    i   the tennis ball   **ii**   the golf ball?

  b What was the highest recorded bounce of
    i   the tennis ball   **ii**   the golf ball?

  c What was the mode for each?

  d Use the mode to help you copy and complete the following:
    'The tennis ball typically bounces ... than the golf ball.'

2 Ishbel went on holiday to Amsterdam.
She counted the different types of vehicles by the number of their wheels (two wheels, three wheels, four wheels, etc.).
She repeated the survey back home.

Amsterdam

No. of wheels	2	3	4	6	8
No. of vehicles	43	3	12	2	0

Glasgow

No. of wheels	2	3	4	6	8
No. of vehicles	5	2	30	5	18

a i How many vehicles did she count in Amsterdam?
  ii How many were bikes?

b i How many vehicles did she count in Glasgow?
  ii How many were bikes?

c Compare the numbers of larger 8-wheeled vehicles she noted in each city.

3 Dave lived in Edinburgh and Pat in Barcelona.
They kept in touch on the internet and one day compared surveys on eye colour.

Edinburgh

Eye colour	blue	green	brown	grey
No. of people	41	28	14	17

Barcelona

Eye colour	blue	green	brown	grey
No. of people	7	9	72	12

a What is the modal eye colour for Edinburgh?

b What is the modal colour for Barcelona?

c How many people were surveyed in each city?

◄◄ RECAP

Pictograms
You should be able to:
- read a pictogram
- make a pictogram, using icons to stand for more than one object.

Example   If we use an anchor, ⚓, to represent 5 boats in a harbour, then we

could represent 4 by ⚓ , 3 by ⚓ , 2 by ⚓ and 1 by |

Bar graphs
You should be able to:
- read a bar graph
- make a bar graph

and use bar graphs to compare sets of data.

Example
This bar graph shows the colours used by a painter over 2 weeks.

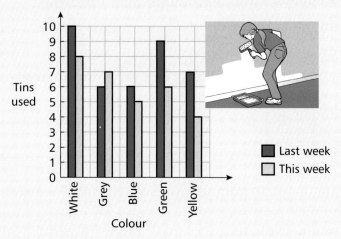

Line graphs
You should be able to:
- read a line graph
- make a line graph

and use line graphs to discuss **trends**.

You should also be able to:
- spot simple trends by looking at a table of values
- spot trends and connections using a **scatter graph**
- compare data sets by using **frequency tables**, especially by comparing **modes**.

1 The pictogram shows the number of flights from some Scottish airports one morning.

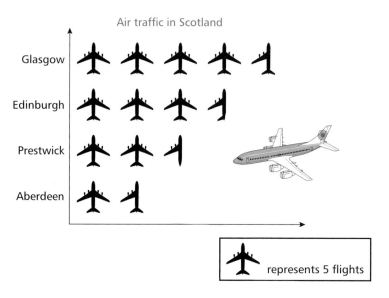

Air traffic in Scotland

✈ represents 5 flights

   a How many flights took off from Glasgow?

   b How many more took off from Edinburgh than Prestwick?

   c How many flights were noted in total?

2 Martin counts the number of windmills in a set of local windfarms.

Windfarm	A	B	C	D
Windmills	10	20	17	12

Copy and complete the pictogram to represent the table using the icon to represent 5 windmills. (Note that will represent 4 mills and 3.)

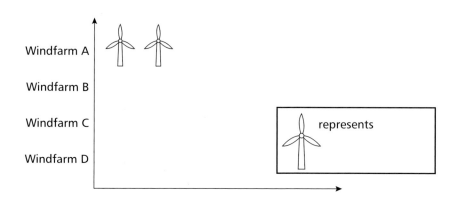

represents

3 Sheila counts the weeds in her vegetable patch.
The bar graph shows her findings.

a How many dandelions did she count?

b How many more buttercups than clover were counted?

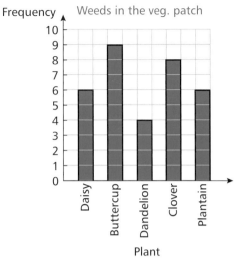

c After some weedkiller is put on, the situation is this:

Plant	Frequency
Daisy	5
Buttercup	3
Dandelion	3
Clover	2
Plantain	4

i Copy the given bar graph.
ii Add the new data in a different colour.

d How many daisies died?

e Which plant was most affected by the weedkiller?

4 Deirdre drops a ball and Allan measures the rebound.

a How far did it bounce when it was released from 120 cm?

b Describe the trend by completing this sentence:
'The higher the release height, the … the ball will bounce.'

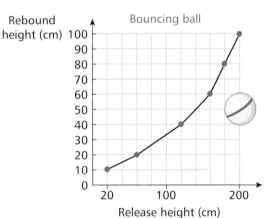

5 Mike sails away from a lighthouse.
He measures the brightness of the beam as he goes.

Distance (km)	1	2	3	4	5
Brightness (units)	250	75	25	15	0

REVISE

163

a Copy and complete the line graph below.

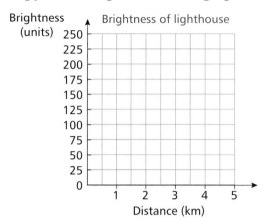

b Complete: 'The further you are from the lighthouse, the … it appears.'

6 Lynn counted the butterflies in her garden over several weeks.
She noted how many weeks had passed since the start of August.

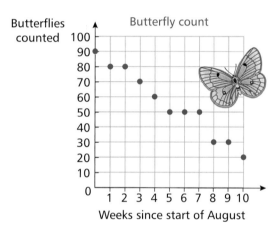

a How many butterflies were spotted on
   i the third week
   ii the seventh week?

b Describe the trend.

7 A passage is chosen from a children's book. The length of each word is noted. A second book aimed at adults is picked and the same survey is done.

Children's book

Number of letters	4	5	6	7	8+
Number of words	67	54	45	24	10

Adult's book

Number of letters	4	5	6	7	8+
Number of words	36	45	65	39	15

a What is the most common length of word in
   i the children's
   ii the adult's book?

b How many more words of 4 or 5 letters are there in the children's book compared with the adult book?

c How many words were looked at in each book?

11 Three dimensions

Builders work from drawings.
The drawings are on a 2-D (flat) surface.

The builder has to build 3-D (solid) objects from
the information given on the drawings.

Being able to understand things in 2-D and then
transform them into 3-D is an important skill for
builders.

1 Review

◀◀ Exercise 1.1

1 Look at this square.

   a What do the boxes in the corners tell us?

   b What do the marks on the sides tell us?

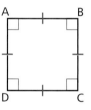

2 Here is a rectangle.

   a Why do PS and QR have one mark while PQ and
SR have two marks?

   b What can you say about a square that you can't
say about a rectangle?

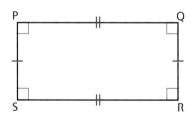

3 Here are two triangles. One is an isosceles triangle.
The other is an equilateral triangle.

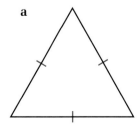   

Which one is the equilateral triangle?

4 A radius and diameter have been drawn on a circle.
Name: **a** the radius
        b the diameter.

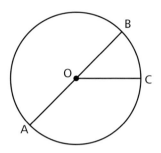

5 a Name this rectangle.

  b Name the side at the top of the rectangle.

  c Name a side equal to PS.

  d Name the diagonals.

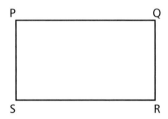

2 Faces, edges and vertices

A **face** is an outside surface of a 3-D shape.

An **edge** is where two faces meet.

A **vertex** is where edges meet.

We say one vertex but two or more **vertices**.

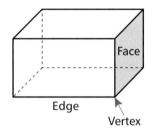

Exercise 2.1

1 Each angle on each face is a right angle.

  a What do we call this 3-D shape?

  b Why are some of the edges shown as dotted lines?

  c How many faces does the shape have?

  d How many edges does it have?

  e How many vertices does it have?

  f What shape is each face?

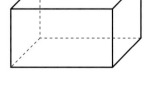

2 Each angle on each face is a right angle.
All the edges are the same length.

  a What kind of 3-D solid is it?

  b What 2-D shape is each face?

  c How many faces does the shape have?

  d How many edges does it have?

  e How many vertices does it have?

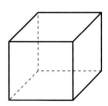

3 ABCD is the nearest face of this cuboid.

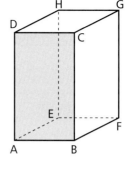

   a AD is the left edge of this face.
      Name the right edge.

   b Name the vertex (corner) which is at the top left of the nearest face.

   c Name the top face.

   d Name the nearest edge of the top face.

   e Name three edges which are parallel to BF.

4 This trinket box is a **triangular prism**.

   a How many of its faces are triangles?

   b How many of its faces are rectangles?

   c How many edges does it have?

   d How many vertices does it have?

5 A paperweight is in the shape of a square-based pyramid. (The base of the pyramid is a square.)

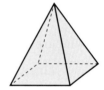

   a What shape are the other faces?

   b How many faces are there altogether?

   c How many edges does it have?

   d How many vertices?

6 This candle is a cylinder.

   a What shape is the top face and the bottom face?

   b How many faces does the cylinder have in total?

7 This party hat is in the shape of a cone.

   a What shape is the base (bottom face)?

   b How many faces does it have?

   c How many edges does it have?

   d How many vertices does it have?

8 A ball has no edges and no vertices and only one face.
This face goes all the way round it.
What is the mathematical name for a ball-shaped object?

9 Copy and complete this table.

Name of 3-D shape	No. of faces	No. of edges	No. of vertices
cube	6	12	8
cuboid			
triangular prism			
cylinder			
square-based pyramid			
cone			
sphere			

3 Drawing cuboids

Exercise 3.1

1 Follow these instructions to draw a 2-D representation of a cuboid on squared paper.
Use a ruler to make your lines nice and straight.

Step 1:   Draw a rectangle.

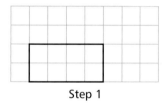
Step 1

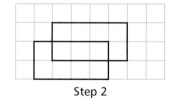
Step 2

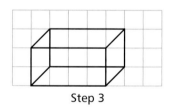
Step 3

Step 2:   Draw a congruent (identical) rectangle moved to the right and up.

Step 3:   Join the corners of the two rectangles as shown.

Step 4:   You must decide which face is nearest you.
Then decide which edges are hidden and draw these as dotted lines.

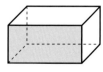 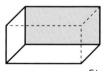
Step 4

Step 5:   You can shade in the faces to make them more solid looking.

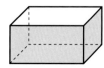 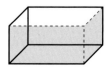
Step 5

2 Use the steps from question **1** to draw a cube with an edge of 3 units (squares).

3 Use the method to help you sketch other shapes, like the ones started here.

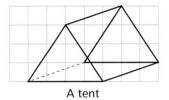

A tent

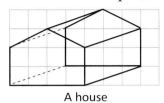

A house

Challenge

Try to make solid letters like this:

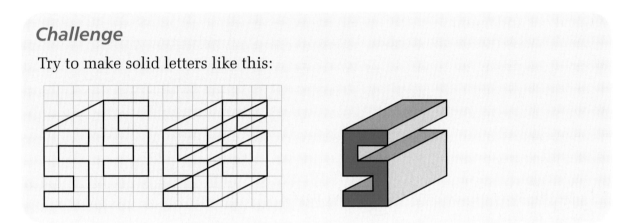

4 The nets of a cube

If you open out a box which is a cube and lay it flat, you get the net of a cube.

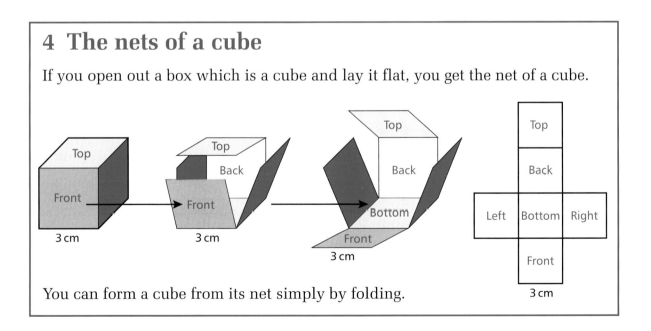

You can form a cube from its net simply by folding.

Exercise 4.1

1 Copy the net of the cube (above) onto centimetre squared paper.
Cut it out and fold along the lines.
See how it folds to make a cube.

2 Same cube but different net! Copy this net onto centimetre squared paper.

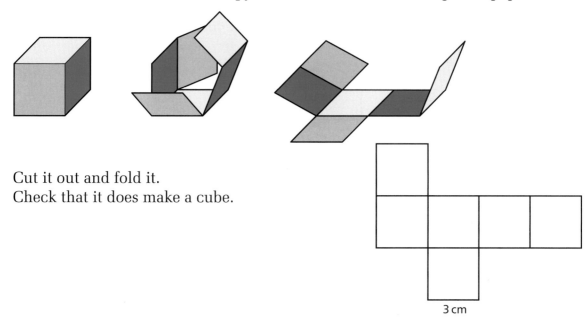

Cut it out and fold it.
Check that it does make a cube.

3 cm

3 Copy this net onto centimetre squared paper.
Cut it out and fold it.

Does it make a cube?

3 cm

4 Here are three nets.
One of them will *not* fold to make
a cube.

Which one?

a   **b**   **c**

5 To find out if a net will fold to make a cube
(without actually folding it), find pairs of
sides which will be opposite each other when
folded.

	A		
B	C	D
E	F	

a Name the face opposite

   i A   **ii** B   **iii** C.

b If A is the top face which is the bottom?
c If B is the front face which is the back?
d If C is the right face which is the left?

The net shown has a top, a bottom, a front, a back, a right face and a left face, so
it will fold to make a cube.

6 Use the same method to work out if these nets make cubes.

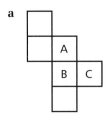 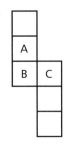 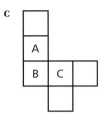

i Sketch the net

ii Mark, if you can, F, the face opposite A.

iii Mark D, the face opposite B.

iv Mark E, the face opposite C.

If this can be done, you have the net of a cube.

Make the nets from paper to check your findings.

Challenge

Note that these four nets are actually the same one.

The net has either been flipped or rotated to make it look different.

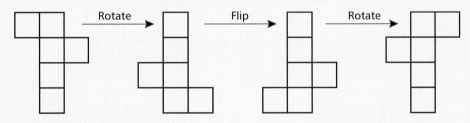

You should be able to see seven nets of cubes in the exercise above.

In total there are eleven possible cube nets that can be made.

Can you find the missing four?

Be careful you don't have two the same.

5 The nets of a cuboid (and other solids)

The nets of a cuboid can be found by opening out old boxes.

Example   A box is 3 cm by 25 cm by 20 cm. Describe its net.

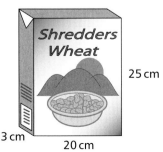

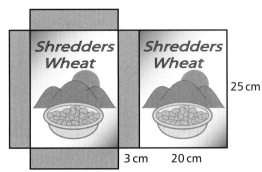

The net has three pairs of **congruent** (identical) faces:
front and back, both 20 cm by 25 cm
top and bottom, both 3 cm by 20 cm
left and right, both 3 cm by 25 cm.

Again, there is more than one possible net for the cuboid.

Exercise 5.1

1 Copy this net of a cuboid onto centimetre
squared paper.
Fold it up to see how it makes a cuboid.

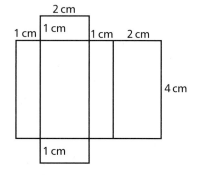

2 This is the net of a cuboid chocolate box.
  a When the box is formed, which side joins to edge
   i AB   **ii** JN
   iii EL   **iv** GF?
  b Point F will meet point A.
   Which other point will meet A?

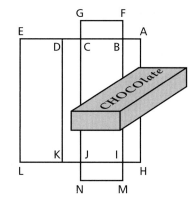

3 Here are three more nets.
Two of them make a cuboid but one doesn't. Which one?

a 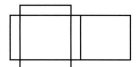   **b** 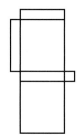   **c**

4 A box of matches is a cuboid
3 cm by 5 cm by 2 cm.
On centimetre squared paper,
draw the net of the box.
Cut it out and fold it to make a cuboid.

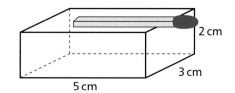

5 A box which holds two packs of cards is a cuboid measuring 2 cm by 8 cm by 12 cm.

  a Make a sketch of a suitable net of the box.

  b The makers want a box that will hold just one of the packs.
  Sketch a suitable net for this box.

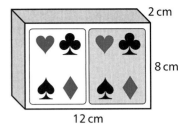

2 cm

8 cm

12 cm

Exercise 5.2

1 This net contains three identical rectangles.

  a Make a net with three identical rectangles.
  (Don't bother with the triangular ends.)
  Cut it out and form a solid.
  Name the type of solid made by this net.

  b What type of triangles are the triangular faces?

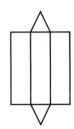

2 This net has four identical isosceles triangles around a square.

  a Name the solid made by this net.

  b Name three points which will join to A when the net is folded to make the solid.

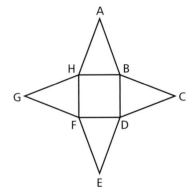

3 The label is removed from this soup tin and laid flat.

  a What type of solid is the tin?

  b What shape is the label?

  c Which of these nets is the net of the soup tin?

  i

  ii

  iii

6 Skeleton models

We can make skeleton models using straws or plastic rods to represent the edges of 3-D solids.

Skeleton models are useful because they let us study the solid as if we can look through it.

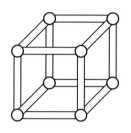

Exercise 6.1

1 Mr Edmund is making a skeleton model of a cubic metre for his classroom.

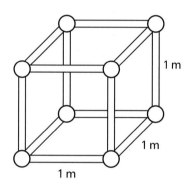

   a How many edges does the cube have?

   b How many 1 metre long rods will he need?

   c What will be the total length of the rods needed to make the model?
     (Remember that all the edges will be the same length.)

   d Special connectors are needed at the vertices (corners).
     How many special connectors will he need?

2 Mr Edmund decides to make a small skeleton model of a cuboid.
The cuboid is to be 20 cm by 12 cm by 8 cm.

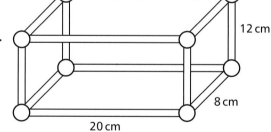

   a How many 20 cm long pieces of straw will he need?

   b How many 12 cm long pieces of straw will he need?

   c How many 8 cm long pieces of straw will he need?

   d What will be the total length of the straws needed to make the model?

   e How many special connectors will be needed for the corners?

3 Lauren is making a cover for one of her patio plants.
She makes a skeleton model of a cube using wooden rods and then covers it with Perspex squares.
The cube has edges of length 50 cm.

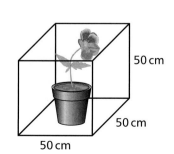

   a How many rods will she need?

   b What total length of rod will she need?

   c How many Perspex squares will she need?

4 Richard is making a run for his rabbit.
He will use wooden rods to make a skeleton
model of a cuboid.
He will then use wire netting to cover it.
The run will be 3 metres by 2 metres by 1 metre.
What is the total length of wooden rods that he
will need?

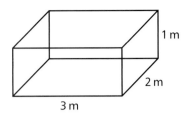

7 3-D problems

Exercise 7.1

1 A box is made to hold three golf balls.
The golf balls have a diameter of 4·3 cm
  a What is the length of the box?
  b What is the breadth of the box?
  c What is the height of the box?

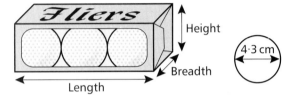

2 Tins of beans are kept in a box.
The box is a cuboid.
The tins are in four rows.
There are five tins in each row.
The tins are 11 cm tall and have
a diameter of 7 cm.
  a What is the length of the box?
  b What is the breadth of the box?
  c What is the height of the box?

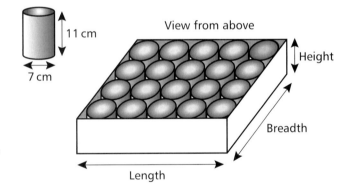

3 Jenny is moving home and she needs to put her
computer games into a box.
Each game measures 20 cm by 14 cm by 2 cm.
The box measures 40 cm by 28 cm by 20 cm.
  a How many games could be laid into the
  bottom of the box?
  b How many layers of games could be fitted in?
  c How many games could be fitted into the
  box in total?

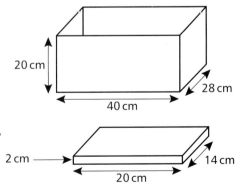

◀◀ RECAP

A **face** is an outside surface of a solid.

An **edge** is where two faces meet.

A **vertex** (corner) is where two or more edges meet.

The dotted lines show hidden edges.

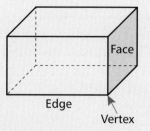

A **net** of a solid is what the solid would look like if it was unfolded and all its faces laid flat.

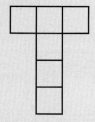

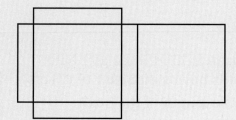

This is a net of a cube.　　　　This is a net of a cuboid.

The same solid can have several different nets.
These are all nets of the same cube.

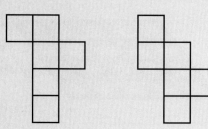 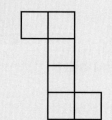 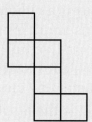

A cube has 11 different nets.

Skeleton models are good for examining solids.

We use straws or rods to represent the edges of the solids.

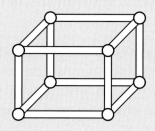

1 a A solid has six square faces which are all the same size.
  Is it a cube or a cuboid?

  b A solid has six faces which are all rectangles. The top and the bottom faces are the same as each other. The front and the back are the same as each other. The right side and the left side are the same as each other.
  Name the solid.

2 ABCD is the top face of this cuboid.

  a Name the bottom face.

  b Name the edge which is at the bottom of the right-hand face.

  c Name the vertex which is at the top right corner of the front face.

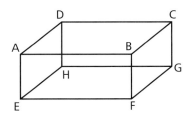

3 Which of these can be cut out and folded to make a cube?

a 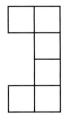   **b** 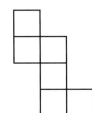   **c** 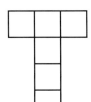   **d**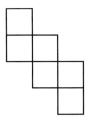

4 Match each solid with a net.

a   **b**   **i** 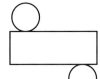   **ii**    **iii**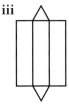

c

5 A ladybird has landed on this skeleton model of a cuboid.
  It wants to go from A to Y.
  It must walk along the edges.
  What is the shortest distance it must walk to get from A to Y?

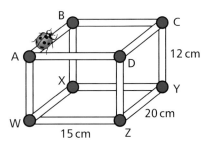

12 Logic diagrams

Instructions need to be carried out in the correct order.

If they are done in the wrong order then the result may not be what you expected!

1 Review

◀◀ Exercise 1.1

1 The table gives the cost of cinema tickets. How much does it cost for:
   a an adult at the weekend
   b a child on Tuesday
   c an adult with a child on Saturday?

	Weekdays	Weekends
Adult:	£5	£7
Child:	£2	£4

2 Find the missing numbers in each diagram.

a (2)—Add 4 > Double > (?)

b

7	9	?	?

+2   +2   +2

c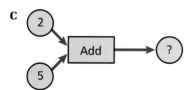

d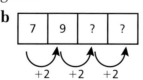

3 How to boil water:

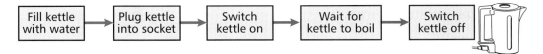

a What is the first instruction?

b What is the last instruction?

c What do you do after switching the kettle on?

d What do you do before plugging the kettle into the socket?

4 Tower instructions:

   i Put part A on top of part B.   **ii** Put part C on top of part A.

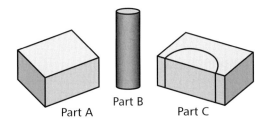

Which of these towers is correct?

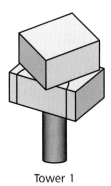
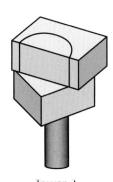
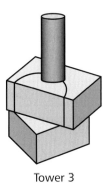

Tower 1   Tower 2   Tower 3

2 Tree diagrams

Iain is buying a new sports top.

He is asked:
- Round-neck or V-neck?
- Plain or striped?
- Long or short sleeves?

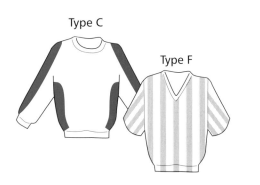

Type C

Type F

Here are all the possibilities shown in a **tree diagram**.
Each decision creates a new branch.

Round — Design? — Striped — Sleeves? — Long — Type A / Short — Type B
Round — Design? — Plain — Sleeves? — Long — Type C / Short — Type D
Neck? — V-neck — Design? — Striped — Sleeves? — Long — Type E / Short — Type F
Neck? — V-neck — Design? — Plain — Sleeves? — Long — Type G / Short — Type H

Type C, for example, is a round-necked plain long-sleeved top.
Another example is Type F. This is a V-necked striped short-sleeved top.

Exercise 2.1

1 Two coins, a 20 pence and a 10 pence, are tossed.

This tree diagram shows all the possible outcomes.

Copy and complete the table using the tree diagram.

Head — Head — Result A / Tail — Result B
Tail — Head — Result C / Tail — Result D

	20 pence	10 pence
Result A:		
Result B:		
Result C:	Tail	Head
Result D:		

2 A cinema has three types of seats: front stalls, back stalls and balcony.

Here is a tree diagram showing all the different tickets.

Use the tree diagram to help you complete this table.

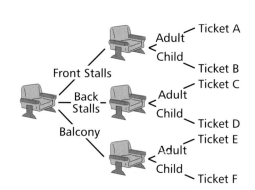

Front Stalls — Adult — Ticket A / Child — Ticket B
Back Stalls — Adult — Ticket C / Child — Ticket D
Balcony — Adult — Ticket E / Child — Ticket F

	Description
Ticket A:	Front stalls, adult
Ticket B:	
Ticket C:	
Ticket D:	Back stalls, child
Ticket E:	
Ticket F:	

3 Look back to the tree diagram of sports tops at the start of this section.
Type C was described as a round-necked, plain, long-sleeved top.
Write descriptions for:

a Type A   **b** Type B   **c** Type D

d Type E   **e** Type F   **f** Type G

4 Pete has gold coins (£2 and £1), silver coins (10p and 5p) and bronze coins (2p and 1p).
He chooses one of each type.
What are the possible values of the three coins he chooses?

Complete this tree diagram to find out.

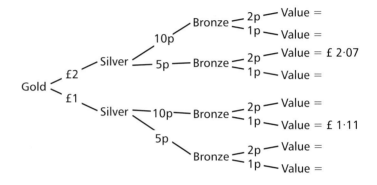

5 A £1 coin, a 50p coin and a 10p coin are tossed.

This tree diagram shows all the possible things that can happen.

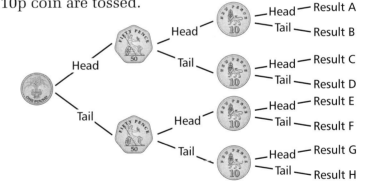

Copy and complete this table using the tree diagram to help you.

	£1 coin	50p coin	10p coin
Result A:			
Result B:			
Result C:			
Result D:	Head	Tail	Tail
Result E:			
Result F:			
Result G:			
Result H:			

6 Traffic lights can be in three states. They can be showing:

- red   (no cars can go)
- amber   (cars might go)
- green   (cars can go)

There are two sets of lights at a crossroads.
One controls the east/west traffic. The other controls the north/south traffic.
The diagram shows all the different states the lights can be in.

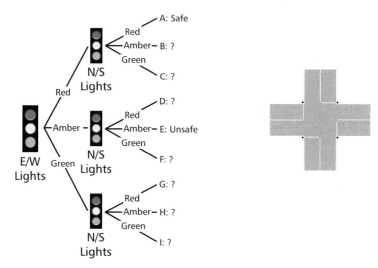

a Why is possibility A **safe**?

b Why is possibility E **unsafe**?

c Decide whether possibilities B, C, D, F, G, H and I are **safe** or **unsafe**.

Challenge

Which coin is it?
Can you identify each of the coins A to H?

3 Time schedules

Bryn and Pat are landscape gardeners.
There are four jobs needing done.

Job 1   1 person   2 hours   Grass cutting
Job 2   2 people   1 hour   Erecting shed
Job 3   1 person   1 hour   Planting
Job 4   1 person   1 hour   Digging (Note: **after job 2**)

Here is Bryn's plan for their time schedule.

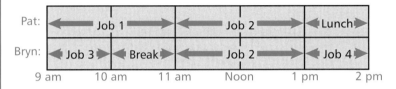

Here is Pat's plan.

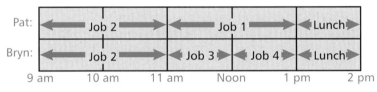

Exercise 3.1

1 Which of the two plans above do you think is better? Explain why.

2 Iain has two jobs to do.

Job 1   1 worker   2 hours   Clean floors
Job 2   1 worker   1 hour   Clean windows

He thinks up two possible plans.

a When does Iain finish Job 1 in   **i** plan A   **ii** plan B?

b In which plan are the windows cleaned first?

c What time have the floors been cleaned in   **i** plan A   **ii** plan B?

d Make a plan for Iain working with Sandy to do the two jobs starting at 9 am.

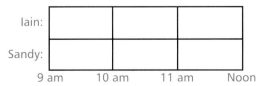

3 Here is another plan of a time schedule.

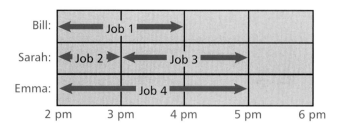

a When does Bill finish job 1?

b Which job does Emma do?

c Who finishes work first?

d Which job is the   **i** longest   **ii** shortest?

e What are they each doing at   **i** 2.30 pm   **ii** 3.30 pm   **iii** 4.30 pm?

f Emma does not turn up for work.
Make a plan to share the four jobs fairly between Bill and Sarah.

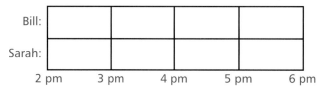

4 David and Craig are car mechanics.
They work from 9 am until 5 pm with 1 hour for lunch.

a Here are the jobs they must do on Monday:

Job 1	2 mechanics	3 hours	(Note: **pm only**)
Job 2	2 mechanics	2 hours	(Note: **am only**)
Job 3	1 mechanic	2 hours	
Job 4	1 mechanic	1 hour	(Note: **am only**)

Create a time schedule for them.

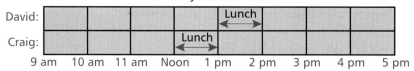

Monday's time schedule

b On Tuesday, David and Craig have five jobs to complete.

Job 1	1 mechanic	4 hours	(Note: **pm only**)
Job 2	2 mechanics	2 hours	
Job 3	1 mechanic	3 hours	
Job 4	1 mechanic	2 hour	
Job 5	1 mechanic	1 hour	

Make a plan showing when they should do the jobs.

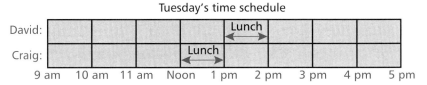

Tuesday's time schedule

| David: | | | | | Lunch | | | | |
| Craig: | | | | Lunch | | | | | |

9 am 10 am 11 am Noon 1 pm 2 pm 3 pm 4 pm 5 pm

4 Flow diagrams

A **flow diagram** can be used to show the order of instructions.
Sometimes you make a decision before you do something.
Flow diagrams have decision boxes to show this. They are diamond shaped.
The two exits from the decision box should be labelled 'yes' and 'no'.
The diagram should have a title explaining its purpose.

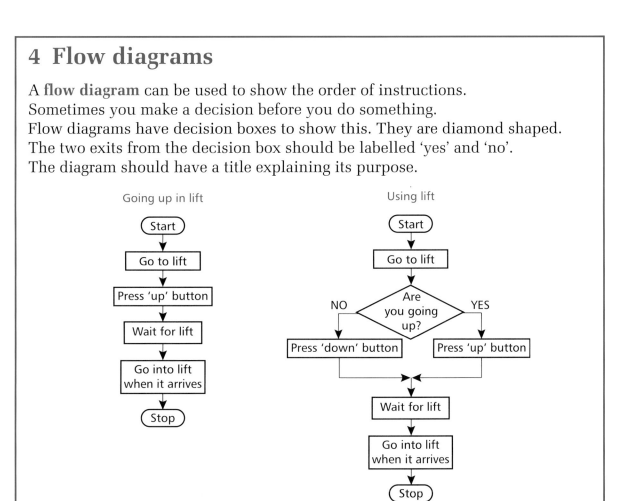

Exercise 4.1

1 Sort the instructions into the right order and suggest a title.

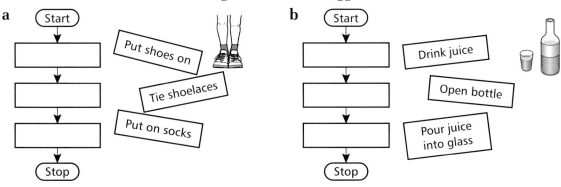

a Start
Put shoes on
Tie shoelaces
Put on socks
Stop

b Start
Drink juice
Open bottle
Pour juice into glass
Stop

c

Start

Watch DVD

Put disk into player

Press play button

Switch on player

Stop

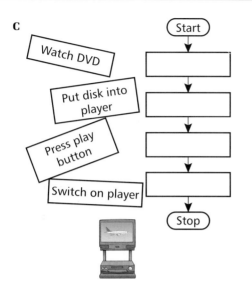

d

Start

Break egg into pan

Pour oil into pan

Wait for oil to heat

Put pan on hotplate

Turn on hotplate

Stop

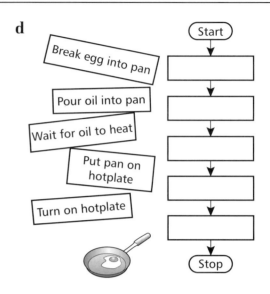

2 a Use this flow diagram to find the cost of hiring a van for:
  i 1 day
  ii 3 days
  iii a full week.

  b i Ellie hired a van for Monday and Tuesday. How much did this cost her?
  ii She hired it again for Thursday, Friday, Saturday and Sunday. How much did this cost?
  iii Would she have been cheaper hiring it for a full week? Explain.

Van hire charges

Start

Multiply £15 by the number of days

Add £25

Stop

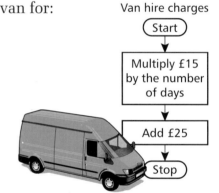

3 Use this flowchart to find the cost of hiring:
  a a special mountain bike for 3 hours
  b a normal bike for 2 hours
  c two special mountain bikes for 5 hours.

Bike hire charges

Start

Are you hiring a special mountain bike?

NO — Multiply £3 by the number of hours

YES — Multiply £5 by the number of hours

Stop

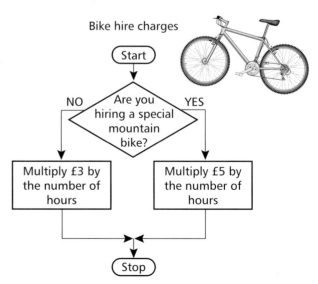

4 Use the flowchart to answer the following questions.

a What question do you ask after putting the alarm off?

b What happens if you are not ready to get up?

c If you reset the alarm what do you do next?

d What happens just before you decide whether or not you are getting up?

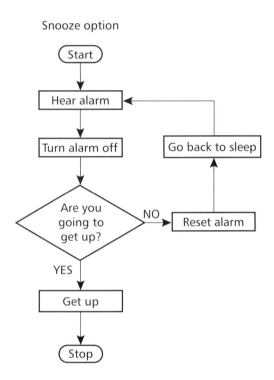

5 This flowchart helps you change between two temperature scales.

Change:

a 10 °C to °F

b 41 °F to °C

c 59 °F to °C

d 25 °C to °F

e 212 °F to °C

f 100 °C to °F

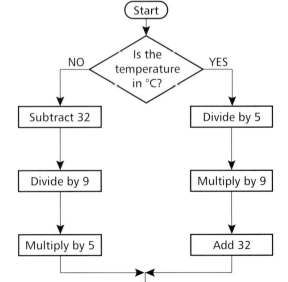

6 Use the flow diagram to find the cost of:

   a 20 litres of normal

   b 15 litres of super

   c 26 litres of super

   d 50 litres of normal

Normal: 86p per litre

Super: 95p per litre

10% discount on purchases greater than £20

Petrol charges

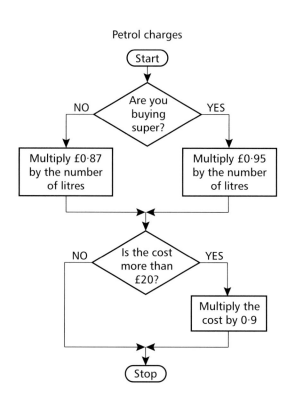

Challenge

Natalie works at a garden centre. She is in charge of the greenhouses.

The temperature must always be between 18 °C and 25 °C.

If it is too cold she turns the heaters on.

If it is too hot she switches the heaters off.

Design a flow diagram giving her clear instructions.

◄◄ RECAP

Tree diagrams

You should be able to use a tree diagram to help you sort out the different things that can happen in a situation.

Example
Ailsa can travel to work by bus, upstairs or downstairs, or by train, 1st class or 2nd class.
There are four different possibilities:

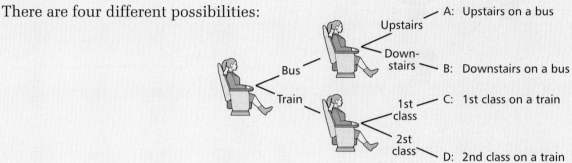

A: Upstairs on a bus
B: Downstairs on a bus
C: 1st class on a train
D: 2nd class on a train

Time schedules

You should be able to make a plan for sharing work jobs.

Example

Job 1	2 people	1 hour
Job 2	1 person	2 hours
Job 3	1 person	1 hour
Job 4	1 person	1 hour

One possible plan for Elaine and Sarah to share these jobs is:

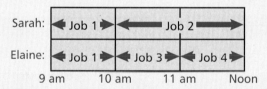

Flow diagrams

You should be able to follow the instructions in a flow diagram.
You should also be able to turn instructions into a flow diagram to make them clearer.

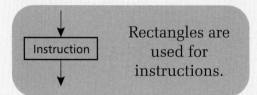

Rectangles are used for instructions.

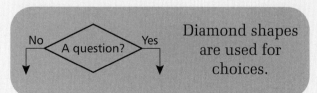

Diamond shapes are used for choices.

1 A shop sells T-shirts in three different sizes: small, medium and large.
They come in two colours: white and black.
Use the tree diagram to help you complete the table.

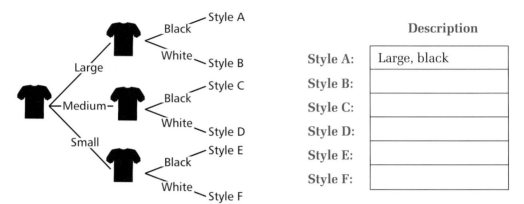

Style	Description
Style A:	Large, black
Style B:	
Style C:	
Style D:	
Style E:	
Style F:	

2 Katie and Grace are decorators.
On Saturday they work from 8 am until 2 pm.
They have four jobs to complete.

Job 1	2 decorators	1 hour
Job 2	1 decorator	2 hours
Job 3	1 decorator	4 hours
Job 4	1 decorator	2 hours	(Note: **am only**)

Create a time schedule for them.

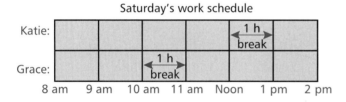

3 Mike likes to download tunes from the web.
Use the flow diagram to find the cost of downloading:

a 2 songs

b 20 songs

c an album of 12 songs

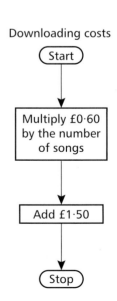

4 Use the rough conversion flow diagram to change:

   a 10 feet to metres

   b 3 metres to feet

   c 12 metres to feet

   d 5 feet to metres

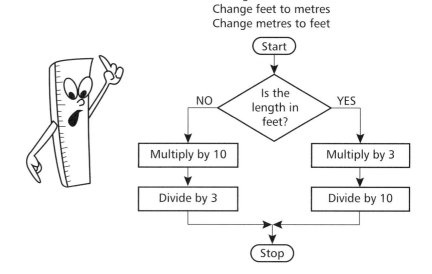

Rough conversions
Change feet to metres
Change metres to feet

REVISE

13 Chapter revision

Revising Chapter 1 Earnings

1 Ali's weekly wage is £248·50.
How much does he earn in:

a 5   **b** 10 weeks?

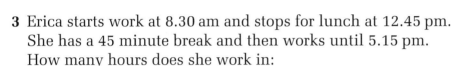

2 In 10 months Vera earns a total of £24 628.
Calculate how much she earns each month.

3 Erica starts work at 8.30 am and stops for lunch at 12.45 pm.
She has a 45 minute break and then works until 5.15 pm.
How many hours does she work in:

a a day   **b** a 5 day week?

4 Andy's hourly rate of pay is £7·85.
How much does he earn in:

a a day when he works for 7 hours   **b** a 35 hour week?

5 Moira's basic rate of pay is £9·47 per hour.
Overtime on Sundays is paid at double time.
How much does she earn for working:

a 1 hour on a Sunday   **b** 4 hours on a Sunday?

6 Mr Cartwright, a car salesman, is paid a commission of 3% on his sales.
Calculate the commission on a car he sells for

a £800   **b** £6500.

7 Sarah sews glove puppets.
She is paid £13·50 for each puppet she makes.
How much does she earn when she makes

a 6   **b** 30 puppets?

8 This column of George's weekly payslip
shows his gross pay and deductions.
Calculate his take-home pay.

Gross pay £384·08
Deductions £132·96
Take-home pay

9 This is Mrs Cooke's monthly payslip.

Gross pay A. Cooke	Employee number 204	NI number YM 305791B	Month number 4
Basic pay £1840·49	Overtime £74·27	Bonus £28·00	Gross pay
Income Tax £416·36	NI £196·68	Pension £85·64	Total deductions
			Takle-home pay

Calculate her: **a** gross pay
b total deductions
c take-home pay

10 Pierre makes a regular saving out of his wages of £12·50 each week.

  a How much does he save in 1 year (52 weeks)?

  b How long does it take him to save £200?

Revising Chapter 2 Proportion

1 Cinema tickets cost £3·50 per person.
How much does it cost for 3 tickets?

2 Alan pays £300 to hire a car for 5 days.
How much does the car hire cost per day?

3 Kim cycles 16 laps of the track in 8 minutes.

  a How many laps did she cycle in 1 minute?

  b How many laps did she cycle in 6 minutes?

4 A book of six first class stamps costs £1·80.
How much does a book of ten first class stamps cost?

5

  a Calculate the cost of one bar for each pack.

  b Which is the better buy?

6 Linda mixes 1 part lemon juice with 3 parts lime juice to make a lemon and lime drink.

  a How many millilitres of lime juice should she mix with 50 ml of lemon?

  b How many millilitres of lemon juice should she mix with 600 ml of lime?

REVISE

7 This is Abdul's recipe for chapatis. It makes ten chapatis.

Wholemeal flour   200 g
Plain white flour   100 g
Salt   2 ml
Warm water   160 ml
Butter   10 g

How much of each ingredient is needed to make:

a 5   **b** 15 chapatis?

8 A cyclist's map is made using a scale of 1 cm to 2 km.

  a On the map the distance between two stages is 15 cm.
    How far is the actual distance?

  b One particular stretch of road is 8 km long.
    What distance shows this on the map?

9 The graph helps you to change
between kilograms and pounds.

  a Change 8 kilograms to pounds.

  b Change 11 pounds to kilograms.

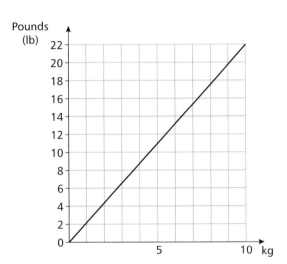

Revising Chapter 3 Perimeter and area

1 a Write down the length of
    i AB
    ii AF.

  b Calculate the perimeter of the shape.

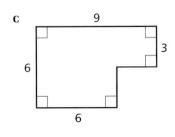

2 Find the perimeter of each of these shapes.
All measurements are in centimetres.

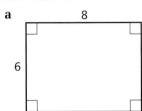

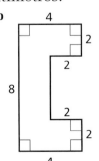

3 The Hotel Coronas has a swimming pool
shaped like the one in the diagram.
Each small square represents an area of 10 m².
Estimate the area of the swimming pool.

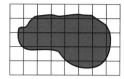

4 Calculate the area of each of these rectangles.

a

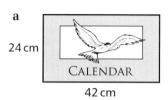

24 cm

42 cm

b

19 cm

14 cm

c

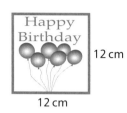

12 cm

12 cm

5 Which picture has:

a the greater area

b the greater perimeter?

Show all your working.

i

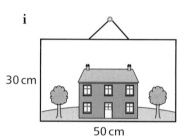

30 cm

50 cm

ii

60 cm

20 cm

6 a

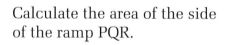

P

3 m

Q   18 m   R

Calculate the area of the side
of the ramp PQR.

b

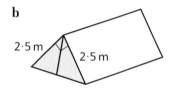

2·5 m

2·5 m

The shaded face of the tent is a right-
angled triangle.
Calculate its area.

7 6 m² of wrapping paper cost £2·40.
What is the cost of 8 m² of the wrapping paper?

Revising Chapter 4   Volume and weight

1 Which unit would you use to weigh:

a a packet of crisps

b an aeroplane

c the weight lifted by a weightlifter?

REVISE

2 Calculate the volume of each cuboid candle.

a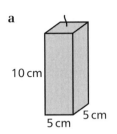

b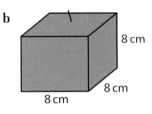

3 a Write down the weight of each parcel.

b What is the difference in weight between the two parcels?

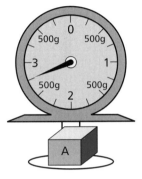 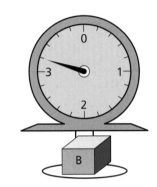

4 Mrs Baker buys a 3 kilogram bag of flour.
She uses 650 grams of it to bake a cake.
What weight of flour does she have left?
Give your answer in **a** grams **b** kilograms.

5 What volume of water will this fish tank hold?

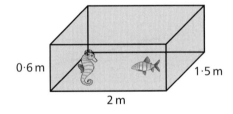

6 Which of these packets of cereal has the greatest volume?

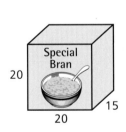 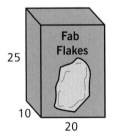 

Explain your answer.
Measurements are in centimetres.

7 Two months ago Bobby weighed 54 kg 400 g.
He now weighs $2\frac{1}{2}$ kilograms less.
What weight is Bobby now?
Give your answer in kilograms and grams.

8 Geri buys these bottles and cans of juice.

850 ml   650 ml   330 ml   1·5 litres

a How much more cola did she buy than cherry?
b How much juice did she buy altogether?
   Give your answer in
   i litres and millilitres   **ii** litres.

Revising Chapter 5   Fractions and percentages

1 Calculate:
   a $\frac{1}{2}$ of 92   **b** $\frac{1}{4}$ of 96   **c** $\frac{1}{5}$ of 115   **d** $\frac{1}{3}$ of 825
   e $\frac{1}{5}$ of 6395   **f** $\frac{1}{2}$ of 82·4   **g** $\frac{1}{6}$ of 36·18   **h** $\frac{1}{7}$ of 85·26

2 Find:
   a $\frac{1}{14}$ of 490   **b** $\frac{1}{24}$ of 1296   **c** $\frac{1}{27}$ of 2295   **d** $\frac{1}{36}$ of 2916

3 What is:
   a 1% of 92   **b** 3% of 77   **c** 4% of 52   **d** 6% of 66?

4 Work out:
   a 10% of 97   **b** 40% of 270   **c** 50% of 860   **d** 20% of 210
   e 25% of 480   **f** $33\frac{1}{3}$% of 390   **g** 25% of 680   **h** $33\frac{1}{3}$% of 570

5 Susan bought a pair of shoes for £52.
   The next week the shop had a sale.
   The same pair of shoes had 10% off the marked price.
   How much did they cost in the sale?

6 There are 180 pupils in the first year at Clinton High.
   Next year there will be 15% fewer.
   How many pupils will there be in next year's first year?

REVISE

7 John needed to save £180 for a new guitar.
After a month, he had saved $\frac{1}{12}$ of this.
How much did he still need to save?

8 A plane took 2 hours 40 minutes to get to its destination flying against the wind.
On the return flight, the time taken was 20% less.
How long did the return journey take?

9 Sam drove 420 miles in 8 hours.
Calculate the rate at which he was driving (in miles per hour).

10 After heavy rain, water was pouring over a dam at a rate of 125 cubic metres per second.
How many cubic metres would pass over the dam in a minute?

Revising Chapter 6   Scale drawing

1 These two photographs are similar.
Calculate the height of the larger one.

15 cm

10 cm

6 cm

Height?

2 The diagram shows the position of boats during a race.

  a Which boat was NE of the *Swift*?

  b Which boat was SW of the *Lucky*?

  c The *Swan* is NW of what boat?

  d The *Heron* is SE of what boat?

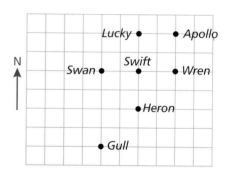

3 A remote controlled plane flew for 35 metres on a bearing of 225°.
It then turned and flew for another 15 metres, this time on a bearing of 160°, before crashing.

a Make an accurate drawing of the plane's route.

b How far was the plane away from its starting point when it crashed?

REVISE

4 Lewis cycled for 6 km.
On a map where 1 cm represents $\frac{1}{2}$ km, what would this distance be represented by?

5 a Give directions for Tom to get from the cinema to the bank.

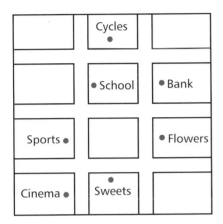

b Give directions for Lynsey to get from the sweet shop to the school.

c Ian turned left out of the school and took the first road on his left.
He then took the first right and first right again.
What was on his left?

d Laura turned left from the cycle shop.
Then she took the first right followed by the next right.
She then took the first left.
What was on her right?

Revising Chapter 7   Patterns and formulae

1

| Pattern 1 | Pattern 2 | Pattern 3 | ? |
| 5 | 9 | ? | ? |

a Draw the next two patterns.

b Find the next three missing numbers.

2 For each number pattern find:
i the rule to get the next number from the previous number
ii the next two numbers.

    a 1, 8, 15, 22, ?, ?

    b 45, 39, 33, 27, ?, ?

    c 3, 14, 25, 36, ?, ?

REVISE

3 Find the missing numbers and rules.

a

Counting numbers	1	2	3	4
Pattern	?	?	?	?

Rule +7

b

Counting numbers	1	2	3	4
Pattern	?	?	?	?

Rule ×4

c

Counting numbers	1	2	3	4
Pattern	11	12	13	14

Rule ?

d

Counting numbers	1	2	3	4
Pattern	7	14	21	28

Rule ?

4 Find the rule that changes:

a 9, 10, 11, 12, … into 1, 2, 3, 4, …   **b** 5, 6, 7, 8, … into 3, 4, 5, 6, …

c 1, 2, 3, 4, … into 20, 40, 60, 80, …   **d** 1, 2, 3, 4, … into 16, 17, 18, 19, …

5 Copy and complete each table.

a

Counting numbers	1	2	3	4	5
Step 1: ×7					
Step 2: −3					

b

Counting numbers	1	2	3	4	5
Step 1: ×3					
Step 2: +8					

6 Find the two-step rule that changes 1, 2, 3, 4, 5, … into:

a 3, 5, 7, 9, 11, …   **b** 5, 12, 19, 26, 33, …   **c** 13, 18, 23, 28, 33, …

7 The tables give the international charges to Malawi made by two phone companies.

Quick Link

Minutes	1	2	3	4	5
Cost (p)	8	11	14	17	20

Wire World

Minutes	1	2	3	4	5
Cost (p)	1	6	11	16	21

a Write down the rule for finding the cost in pence if you know the number of minutes for: **i** Quick Link   **ii** Wire World

b How much does a 10 minute call cost with: **i** Quick Link   **ii** Wire World?

8 These patterns are made with circles.

Pattern 1   Pattern 2   Pattern 3

a Complete the table.

Pattern number	1	2	3	4	5	6		12
Number of circles	6							

b Write down a rule for the number of circles if you know the pattern number.

9 The cost of downloading songs from an internet music site is given by:

cost in pounds = (number of songs) × 0·75 + 0·55

Find the cost of downloading:

a 1 song   **b** 10 songs   **c** an album of 13 songs

Revising Chapter 8   Percentage in money

1 Work these out without using a calculator.

a $33\frac{1}{3}$% of £87   **b** 10% of £680   **c** 50% of £56

d 1% of £3400   **e** 20% of £65   **f** 25% of £52

2 Luke works in a boutique. He has 10% of everything he sells added to his weekly wage. He sells a suit for £150.
How much is added to his wage?

3 Elaine is at the sales. She sees a trouser suit which normally costs £63·90.
In the sale everything is $33\frac{1}{3}$% off.
a How much does she get off?
b How much does she have to pay for the suit?

4 Calculate:

a 8% of £400   **b** 8% of £75   **c** 30% of £620   **d** 85% of £180

e 32% of £950   **f** 12% of £2500   **g** 7% of £180   **h** 14% of £180

i 70% of £65   **j** 16% of £35   **k** 98% of £735   **l** 61% of £26

5 Rory has had a busy week working at a holiday camp.
His boss gives him 17% of his weekly wage as a bonus
because he has worked so hard.
His weekly wage is £165.
How much is his bonus?

6 Work out and round to the nearest penny:
a 19% of £30·10   **b** 42% of £87·20   **c** 81% of £964·30
d 7% of £63·75   **e** 36% of £62·60   **f** 56% of £250·65

7 Pauline invests £350 in a savings account.
The interest rate is 6% p.a.
a How much interest is added to her account after a year?
b How much is now in her account?

8 Blair is due to get a 5% increase in his weekly wage on his eighteenth birthday. His boss decides to give him an 8% increase because his work has improved so much.

At the moment his weekly wage is £148.

a How much is the increase in Blair's weekly wage?

b What is his weekly wage after the increase?

Revising Chapter 9   Tiling and symmetry

1 a Write down the coordinates of:

   i B   **ii** C   **iii** D   **iv** E.

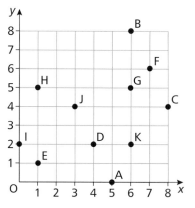

b Name the points at:

   i (6, 5)   **ii** (3, 4)   **iii** (7, 6)

c Name a point on the *x* axis.

d Name a point with an *x* coordinate of 0.

e Name a point on the *y* axis.

f Name a point with a *y* coordinate of 6.

2 Copy and complete these diagrams so that the dotted line is an axis of symmetry.

a
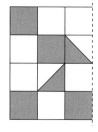

b

c
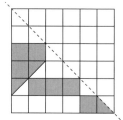

3 Copy these tilings and add at least five more tiles.

a
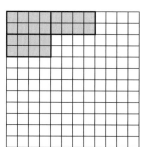

b
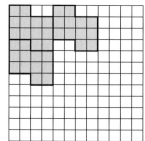

c
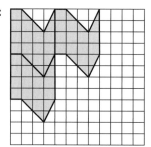

4 Copy this diagram so that the dotted line is an axis of symmetry.
How many axes of symmetry does it have altogether?

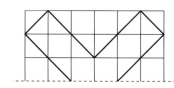

REVISE

Revising Chapter 10 Statistics

1 Leah kept a record of the colours chosen by customers at her nail salon one day.
The pictogram shows her findings.

Nail varnish colour

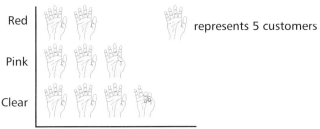

represents 5 customers

a How many chose pink?

b How many more chose clear than red?

c How many customers were noted?

2 Meg noted the rainfall in the first two weeks of April to see which week was wetter. She didn't collect data at the weekend.

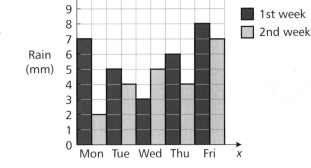

Rain in the first 2 weeks of April

a How much rain fell on the Monday of the first week?

b How much more rain fell on the first Thursday than the second?

c How much rain fell in the first week?

d Which week was wetter?

3 For 10 days after the New Year, Liam measured the outside temperature at noon.
The line graph shows his records.

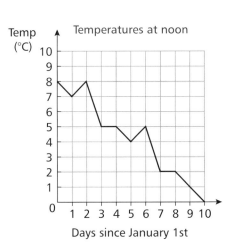

Temperatures at noon

a What was the temperature on
  i the fourth day
  ii the fifth day?

b What is the difference in temperature between the second and the eighth day?

c Describe the trend in the readings.

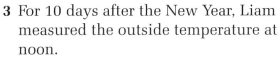

REVISE

4 A teacher logged the progress of 12 students.
In two weeks they sat two mental tests, both
out of 10.
The scatter graph shows the results.
Each dot represents one student.

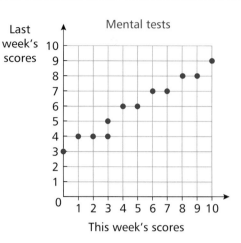

a Last week someone scored 5.
What did he score this week?

b This week someone scored 9.
What did she score last week?

c Consider this week.
What was   i the best   ii the worst score?

d Consider last week.
What was   i the best   ii the worst score?

e Copy and complete:
'The students who did better last week did … this week'

Revising Chapter 11   Three dimensions

1 Name each 3-D solid.

a   b   c   d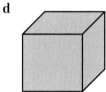

e 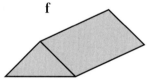   f   g

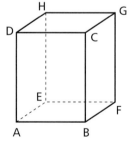

2 CDHG is the top of this cuboid.

a Name the bottom (or base) of the cuboid.

b Name an edge which is parallel to EH.

c Name a face which is congruent (identical) to ABCD.

3 Look at these two nets.
One face on each is
marked with a cross.
Copy the nets, including
the cross.
Put another cross in the
side that will be opposite
the marked side in the solid.

a 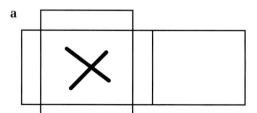   b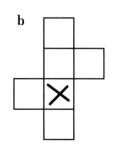

4 Jane is making a skeleton model of a cuboid.
  The cuboid is to be 10 cm by 8 cm by 4 cm.
  a How many rods (or straws) of each length will she need?
  b What is the total length of rod (or straw) required?

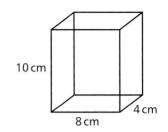

10 cm

8 cm

4 cm

Revising Chapter 12   Logic diagrams

1 Pupils arrive at school either by walking, by car or by bus.
  Some pupils arrive late.
  Use the tree diagram to help you complete the table of possibilities.

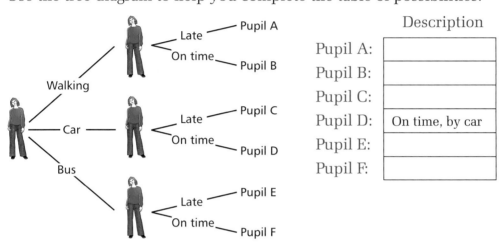

Description

Pupil A:	
Pupil B:	
Pupil C:	
Pupil D:	On time, by car
Pupil E:	
Pupil F:	

2 Julie is sending a letter, a package and a parcel.
  Each could be sent by first class or second class mail.
  The charges are:

	1st class	2nd class
Letter:	30p	21p
Package:	94p	71p
Parcel:	£2·15	£1·75

Using this tree diagram, calculate all the possible costs of her postage.

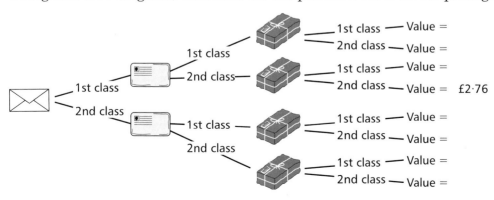

REVISE

3 Ali and Anthea work at the local supermarket in the evening.

Job 1   1 assistant   1 hour   Checkout
Job 2   2 assistants   2 hours   Stock taking
Job 3   1 assistant   2 hours   Shelf stacking

Complete this time schedule for Ali and Anthea to share these three jobs.

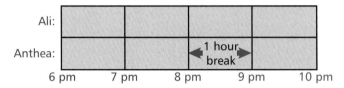

4 Here is another time schedule.

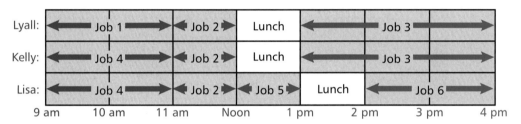

a Who shares job 4?
b Which job requires all three people?
c How many jobs does Lisa complete before lunch?
d Who never does a job on their own?
e What are they each doing at:
    i 9.30 am   **ii** 12.30 pm   **iii** 1.30 pm?

5 The instructions for crossing a road safely are jumbled up.
Sort them into the correct places in the flow diagram.

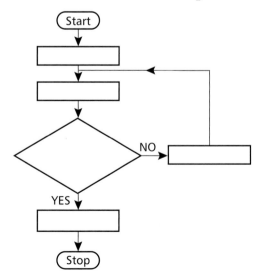

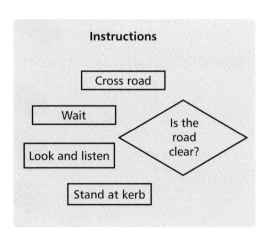

Instructions

Cross road

Wait

Look and listen

Is the road clear?

Stand at kerb

6 Use the flow diagram to find the cost of downloading:

a 5 songs

b an album with 11 songs

c 20 songs from different albums

d 2 complete albums, one with 15 songs, the other with 10 songs.

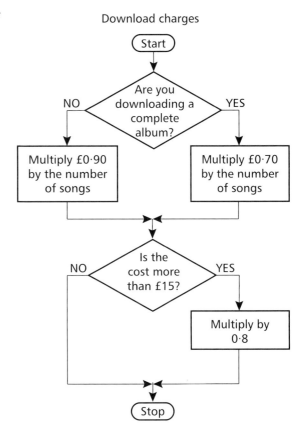

Download charges

REVISE

14 Preparation for assessment

REVISE

Test A Part 1

1 Calculate:

    a 875 + 497   **b** 85 × 6   **c** 255 ÷ 5   **d** 2·58 × 10

2 a Round 62 to the nearest 10.

    b Round 8534 to the nearest 100.

3 Calculate $\frac{1}{4}$ of 892.

4 What is 1% of £24?

5

Distance travelled in km = (speed in km/h) × (number of hours travelled)

Use this formula to find the
distance travelled by a train going
at 90 km/h for 4 hours.

6 Use the flow diagram to find the new speed on each speedometer.

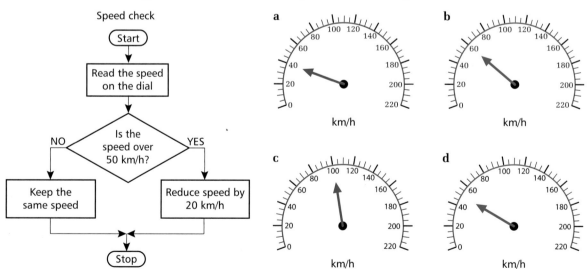

7 Gavin spends 25% of his weekly wage on groceries.
His weekly wage is £148.
How much does he spend on groceries?

8 Copy and complete the diagram so that the dotted line is an axis of symmetry.

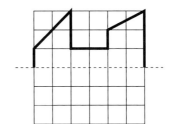

9 Wullie is painting an old set of building bricks for his grand-daughter.
The bricks are cubes and there are 12 bricks in the set.

a How many faces does one brick have?

b How many faces will he need to paint?

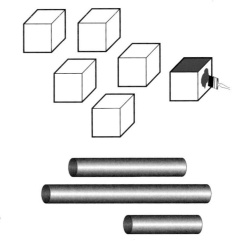

10 A length of plastic pipe is cut into three pieces.
They are 350 mm, 680 mm and 840 mm long.
What is the total length of pipe in

a millimetres   **b** centimetres   **c** metres?

11 A teacher asked her class:
'What type of calculator do you have?'
The graph shows the replies.

a What was the most common type of calculator owned?

b How many pupils were in the class?

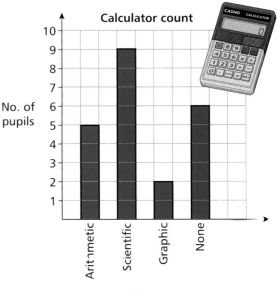

12 Aziz noted the colour showing on 20 sets of traffic lights as he approached them:

green	red	red	green
green	red	amber	green
red	green	red	green
amber	green	red	red
red	red	green	red

Colour	Tally	Frequency
Red		
Amber		
Green		

a Copy and complete the frequency table.

b Which colour is the mode?

Test A Part 2

1 Six filing cabinets cost £744.
How much does one cabinet cost?

2 Calculate:
   a $2\cdot57 + 45\cdot86$ **b** $4\cdot79 \times 4$ **c** $26\cdot5 \div 5$

3 a How heavy is each parcel?

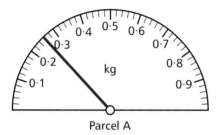
Parcel A

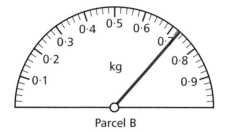
Parcel B

  b What is the difference in their weights?

4 What is the total bill?

Gas bill	
Cost of gas	£58·46
Standing charge	£5·35
VAT	£3·19
Total bill	

5 Carly decides to set herself up as a DJ.
She buys the equipment on hire purchase.
  a Calculate the deposit.
  b What is the total cost of the monthly payments?
  c What is the total HP cost?
  d How much more is the HP cost than the cost price?

BE A DISC JOCKEY
Complete DJ set
Cash price £2500 or
HP terms Deposit 10%
plus 10 monthly payments
of £245

6 Kay is paid £1·50 for each car she cleans.
  a How much does she earn for cleaning five cars?
  b How many cars would she need to clean to earn £12?

7 a Diane saves £20 from her weekly wage towards a holiday.
    How long will it take her to save £500?
  b Mr Parks saves £100 each month.
    How much does he save in one year?

REVISE

8 Opera singer Julio Tenorious pays his agent 7% of his earnings.
Julio was paid £34 000 for three appearances at Glasgow Concert Hall.
How much did he pay his agent?

9 a How much water is in each jug?

b How much more water is in the second jug?

10 Dice are placed in a row as shown.

1 dice on table

5 faces show

2 dice on table

8 faces show

3 dice on table

11 faces show

a Copy and complete the table.

No. of dice	1	2	3	4	5	12
No. of faces showing	5	8	11			

b Write down the rule
for finding the number of faces showing if you know the number of dice.

11 A traffic engineer counted the number of potholes in the road.
He did this for eight stretches of road and found:
5 holes, 8 holes, 12 holes, 4 holes, 2 holes, 9 holes, 11 holes, 5 holes.

a How many holes did he count in total?

b What is the mean number of holes per stretch?

12 A new bus service has started operating.
Each day the number of passengers that catch the bus is counted.
The graph shows the results.

a How many people caught the bus on the fifth day?

b Describe the trend in the number of people catching the bus.

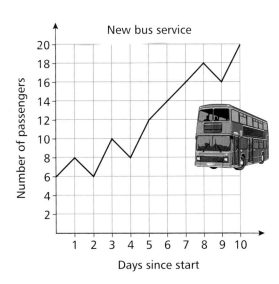

REVISE

Test B Part 1

1 Calculate:

    a $6349 + 267$ **b** 98×4 **c** $765 \div 3$ **d** $83{\cdot}02 \times 100$

2 a Round 86 to the nearest 10.   **b** Round 9797 to the nearest 100.

3 Calculate $\frac{1}{9}$ of 2106.

4 What is 3% of 98?

5 ┃ Distance travelled in km = (speed in km/h) × (number of hours travelled) ┃

Use this formula to find the distance travelled by a cyclist doing 25 km/h for 2 hours.

6 Sort this jumbled flow diagram.
It gives instructions for changing between minutes and seconds.

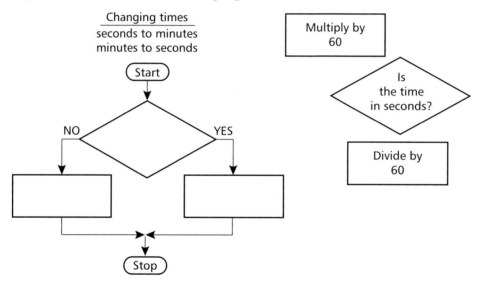

7 Colin is a plumber. He receives a 10% increase in his wages.
His wage was £335 a week before the rise.

    a How much was the increase in his wages?

    b How much is he paid each week after the wage increase?

8 Copy and complete this diagram so that the dotted line is a line of symmetry.

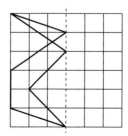

REVISE

9 The net of a cube is shown.

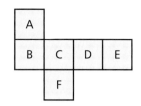

a Face C is the base (bottom) of the cube.
Which face is the top?

b Face F is the front of the cube.
Which face is the back?

10 This graph was produced from a survey into the lunchtime habits of pupils.

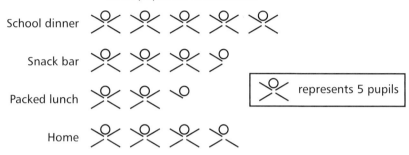

represents 5 pupils

a How many went to the snack bar?

b How many were asked?

c How many did not go home?

11 These two rectangles are similar.
Find the height of the bigger one.

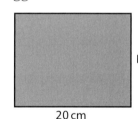

12 cm

16 cm

20 cm

Height?

12 A class of pupils were asked: 'How did you get to school today?'
These are the replies:

walk	walk	bike	walk	bus
walk	bus	car	car	car
car	bike	walk	car	bus
bus	walk	bus	bike	walk

a Copy and complete the frequency table.

b What method of getting to school is the mode?

Method	Tally	Frequency
Bus		
Car		
Walk		
Bike		

REVISE

213

Test B Part 2

1 Mr Jones usually takes 36 minutes to get to work.
His trip took 6 times longer one day due to bad weather.
How long did it take him?

2 Calculate:

   a $97·36 + 86·49$ **b** $13·63 \times 6$ **c** $52·29 \div 7$

3 Sara got two photos of her gymnastic display.
They were similar.
What is the height of the smaller picture?

6 cm

12 cm

4 cm

Height?

4 Jordan bought a car for £6380.
The next year he sold the car for 40% less.
How much did he sell it for?

5 Copy and complete this bill.

```
        Decorating bill

8 rolls of wallpaper at £6·45 per roll    .........
5 metres of border at £4·99 per metre     .........
2 packets of paste at £1·86 per packet    .........

                        Total bill    _____
```

6 Jim works 7 hours and 30 minutes each day
from Monday to Friday.
How many hours is this in total?

7 Phil delivers leaflets. He is paid 50p for every 10 houses.
How much is he paid for delivering 600 leaflets?

REVISE

8 5 metres of electric cable cost £12·50.

   a What is the cost per metre of the cable?

   b How much would 8 metres cost?

9 Terry worked at the Gamekeeper's Arms Hotel.
He was paid £4·00 per hour for washing dishes.
When he started to work as a waiter his pay went up by 12%.

   a By how much did his pay go up?

   b What is he paid as a waiter?

10 a Measure the edges of this shape with a ruler.

   b Calculate the shape's perimeter.

11 a Calculate the volume of the box.

   b A second box, twice as long, twice as broad
and twice as high is made.
Calculate the volume of this second box.

   c How many of the smaller box should be able
to fit inside the larger box?

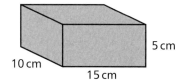

12 Lamp-posts along a street are evenly spaced.
You walk past 2 lamp-posts in 10 m.
You walk past 3 lamp-posts in 20 m.
You walk past 4 lamp-posts in 30 m.

   a Copy and complete the table.

Lamp-posts	2	3	4	5	6	12
Distance walked (m)	10	20	30			

   b What is the rule for the distance walked
when the number of lamp-posts is known?

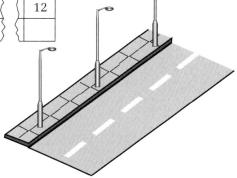

Test C Part 1

1 Calculate:

    a $238 + 3186$ **b** 235×5 **c** $952 \div 4$ **d** $0\cdot03 \times 100$

2 a Round 755 to the nearest 10.

   b Round 34·68 to the nearest whole number.

3 Calculate $\frac{1}{7}$ of 9884.

4 What is 5% of £88?

5 | Distance travelled in km = (speed in km/h) × (number of hours travelled) |

   Use this formula to find the distance travelled by a plane flying at 500 km/h for 6 hours.

6 Sort this jumbled flow diagram.
   It gives instructions for changing between weeks and days.

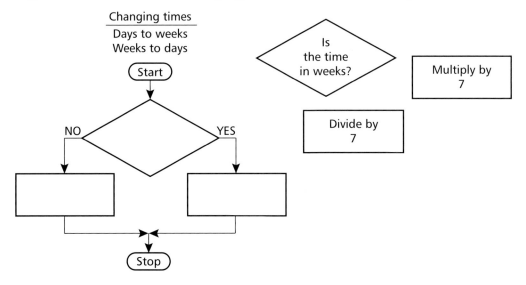

7 Copy and complete the diagram so that the finished picture has an axis of symmetry as shown.

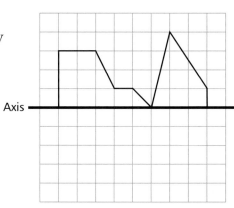

REVISE

8 Name a 3-D shape that has:

  a one face, no vertices and no edges

  b three faces, two edges and no vertices

  c two triangular faces and three rectangular faces.

9 Find the area of this shape.
Each small square represents an area of 1 cm².

10 The doors are similar rectangles.
What is the width of the smaller door?

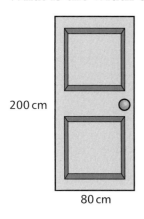

200 cm

150 cm

Width?

80 cm

11 The graph shows the number of customers
counted in a newsagent's, on the hour, right
through the day.

  a How many customers came in when the
shop opened at 7 am?

  b What was the least busy time of day?

  c At what times were four customers
recorded?

12 Ben thought that the dice he was
playing with was not true.
He threw it 20 times with these results:

    1   1   1   2   3

    1   1   4   1   2

    2   5   2   4   3

    2   5   3   3   1

Score	Tally	Frequency
1		
2		
3		
4		
5		
6		

  a Copy and complete the frequency table.

  b Do you think the dice is fair? Comment.

REVISE

Test C Part 2

1 Calculate:

    a $67{\cdot}02 + 59{\cdot}8$ **b** $22{\cdot}46 \times 5$ **c** $204{\cdot}3 \div 6$

2 A car transporter can carry 12 cars at a time.
How many transporters will be needed to move 90 cars?

3 A film at the cinema lasted for 2 hours 30 minutes.
The same film on DVD lasts $\frac{1}{5}$ longer.
How long is the film on DVD?

4 Copy and complete the table for these sales.

Item	Cost price	Selling price	Profit/loss
Toaster	£12·50	£16·49	
Kettle	£15·99		Profit £6·50
Iron		£24·99	Profit £7·25
Carpet sweeper		£76·45	Loss £12·54

5 Calculate:

    a the VAT

    b the total cost of the gas bill.

> ## Gas bill
> | | |
> |---|---|
> | Cost of gas | £42·00 |
> | VAT at 5% | _____ |
> | Total bill | _____ |

6 Graham is paid a basic rate of £9·68 per hour.
Overtime is paid at double time.
How much is he paid for:

    a 1 hour's overtime?

    b 3 hours' overtime?

7 A liner sails 1920 km in 8 days.
At the same speed how far will it sail in 12 days?

REVISE

8 Mark builds up a row of 'houses' using tiles.
As he adds each tile he works out the perimeter of the overall shape.

1 house
perimeter 6 cm

2 houses
perimeter 10 cm

3 houses
perimeter 14 cm

a Copy and complete the table.

No. of houses	1	2	3	4	5	10
Perimeter of shape (cm)	6	10	14			

b What is the rule for working out the perimeter when the number of 'houses' is known?

9 Emma bought two DVDs in a sale.
They usually cost £15 each.

a How much discount did she get?

b How much did she pay for the two DVDs?

SALE
20% off
all DVDs

10 Serena had £349·70 in her bank account.
The interest rate was 4% p.a.

a How much interest did she receive in one year?
(Round off to the nearest penny.)

b How much did she have in her bank account one year later?

11 Mrs Russell has twin babies.
Moya weighs 3·2 kg and Sean weighs 2·9 kg.

a What is the difference in the babies' weights?
Write your answer in grams.

b What is the babies' total weight in grams?

12 Here are the heights of the boys in Jack's class:

1·43 m, 1·34 m, 1·56 m, 1·30 m, 1·37 m, 1·55 m, 1·24 m.

Jack's own height is 1·41 m. He says his height is above average.

a Calculate the average height of the boys in the class (including Jack).

b Comment on Jack's height.

REVISE

Test D Part 1

1 Calculate:

    a $811 - 209$ **b** 1029×7 **c** $1960 \div 8$ **d** $0{\cdot}2 \times 100$

2 a Round 4689 to the nearest 10.

    b Round 99·35 to the nearest whole number.

3 Calculate $\frac{1}{4}$ of 61·04.

4 What is 10% of £366?

5 A magazine changes its looks. The front page is similar but larger.

    a What is the enlargement factor?

    b What is the height of the larger magazine?

6 Here are the training session times for
Pete's football club.

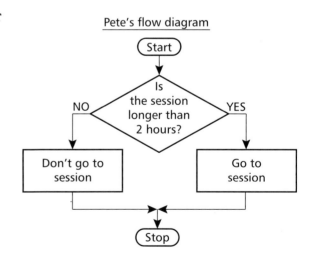

Football training sessions	
Monday	3.30 pm–4.50 pm
Tuesday	3.30 pm–5.00 pm
Wednesday	4.00 pm–6.25 pm
Thursday	6.40 pm–8.10 pm
Friday	5.10 pm–7.15 pm

Pete does or doesn't go to the sessions
depending on the flowchart.

    a How long is the Monday session?

    b Will Pete be going?

    c Which sessions will Pete attend through the week?

7 James invested £530 in a savings account at his bank. The interest rate was 5% p.a.
How much interest would he receive after one year? (Hint: work out 10% first.)

REVISE

8 Copy the diagram and mark in the axes of symmetry.

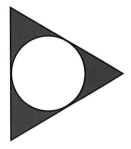

9 Each square on the grid represents 1 cm². What area of the grid has been shaded?

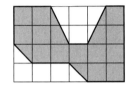

10 The ages of the first 20 pupils to arrive at the school dance were:

15	15	14	16	13
15	16	13	14	12
13	16	15	15	15
13	15	15	16	12

a Copy and complete the frequency table.

b What is the modal age?

Age	Tally	Frequency
12		
13		
14		
15		
16		

11 Here is the net of a cuboid.

a What is the length of AB?

b What is the length of CD?

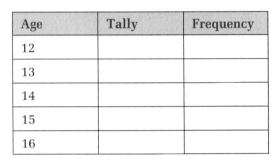

12 The pictogram shows the work done in a day by four brickies.

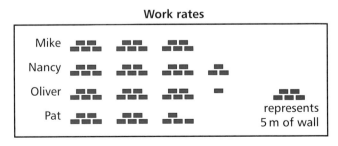

a Who laid the most bricks?

b How long was Oliver's wall?

c What is the difference in length between Nancy's and Pat's work?

REVISE

Test D Part 2

1 Calculate:

    a 75·23 − 23·84   **b** 4·37 × 7   **c** 76·32 ÷ 8

2 Sixteen rolls of wallpaper cost £121·44.
How much does one roll cost?

3 An art dealer buys two paintings. Each one costs £75.

    a He sells one for a profit of 10%.
       Calculate the selling price.

    b He makes a loss of 5% on the other.
       For how much does he sell this painting?

4 Mr Slater decides to buy a new three-piece suite. The cost price is £999.
He takes out a loan to pay for it. He makes 6 monthly payments of £179·95.

    a Calculate the total amount he will pay.

    b How much more than £999 will he pay?

5 a Trixie's weekly wage is £273·50. How much does she earn in 10 weeks?

    b Ralph earns £1840 each month. How much does he earn in 8 months?

6 A 'ship in a bottle' model is made on a scale of 1 cm to 2 metres.

    a The model ship is 15 cm long.
       How long is the actual ship?

    b The mast of the ship is 16 m tall.
       How tall is the mast of the model?

7 Copy and complete each distance table.

    a Average speed is 15 km/h   **b** Average speed is 45 mph

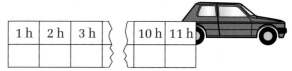

1 h	2 h	3 h		8 h	9 h

1 h	2 h	3 h		10 h	11 h

8 Add five tiles to this tiling.

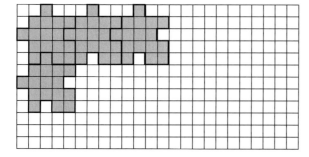

9 a Measure the length and breadth of this rectangle.

  b Calculate the perimeter.

  c Calculate the area.

10 Eight cupfuls of sugar weigh 960 grams.

  a What is the weight of four cupfuls of sugar?

  b What is the weight of ten cupfuls of sugar in kilograms?

11 Helen is making bridges from playing cards.

 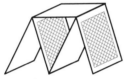 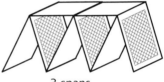

1 span
2 cards

2 spans
5 cards

3 spans
8 cards

  a Copy and complete the table.

Number of spans	1	2	3	4	5
Number of cards	2	5			

  b What is the rule for working out how many cards are needed for a given number of spans?

  c How many spans can be made with a full pack of cards?
(Hint you will only use 50 cards.)

12 Karen notes the length of time each track lasts on her favourite CD:

    3·7 min   3·4 min   3·0 min   2·9 min   3·1 min

    3·6 min   3·6 min   3·7 min   3·6 min

  a What is the modal time?

  b Calculate the mean time a track lasts.

REVISE

Test E Part 1

1 Calculate:

a $1047 - 389$ **b** 3107×8 **c** $6183 \div 9$ **d** $57.72 \div 10$

2 a Round 541 to the nearest 100.

  b Round 77·86 to the nearest whole number.

3 Calculate $\frac{1}{6}$ of 85·5.

4 What is 25% of £644?

5 The diagram shows the positions of mountain rescue teams during a call-out.

  a Which team is NW of Delta team?

  b Which team is Lima NE of?

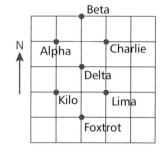

6 Mr Grimes draws a graph to show the cost of running his shower.

  a How much does a 12 minute shower cost?

  b How long does he get for 12p?

7 Yoga class times are shown.

Yoga classes	
Monday	3.40 pm–5.00 pm
Tuesday	3.40 pm–5.10 pm
Wednesday	4.10 pm–6.35 pm
Thursday	6.50 pm–8.20 pm
Friday	5.20 pm–7.25 pm

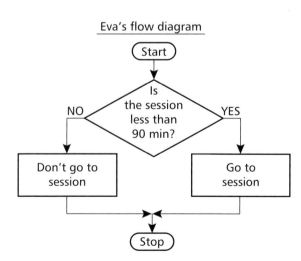

Eva uses the flowchart to decide which classes to go to.

  a How long is the Tuesday session?

  b Will Eva be going?

  c Which sessions will Eva attend through the week?

8 a Write down the coordinates of
    i A   **ii** B   **iii** C.

  b A, B and C are three corners of a square.
    Write down the coordinates of the fourth corner.

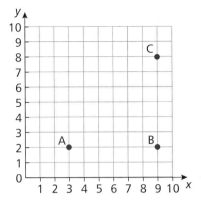

9 Copy the drawing and mark in
the axes of symmetry.
How many axes of symmetry are there?

10 Bruce is making a skeleton model of a cube.
He makes the edges 12 cm long.

  a How many edges does a cube have?

  b What is the total length of the straws that Bruce will need?

11 The bar graph shows how the
audience was made up at a recent
showing of the Western
Where have the buffalo gone?

  a How many male adults were there?

  b How many adults were there?

  c How many more female adults were
there than female children?

  d What was the modal type of audience
member?

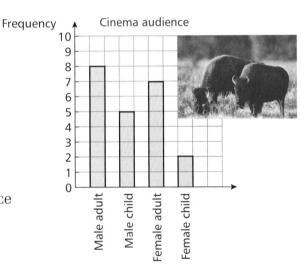

12 People were asked to choose their favourite fruit from apple, orange, pear or
'other'. The results were:

other	orange	apple	orange	apple
orange	orange	apple	apple	other
other	other	apple	pear	pear
apple	apple	orange	other	other

  a Copy and complete the frequency table.

  b What was the overall favourite?

  c Why is it useful to include the
category 'other'?

Fruit	Tally	Frequency
Apple		
Pear		
Orange		
Other		

REVISE

Test E Part 2

1 Calculate:

    a 72·58 − 36·75   **b** 19·54 × 8   **c** 45·81 ÷ 9

2 Alistair flies 54 km in a glider on a bearing of 056°.
    Turning to follow a bearing of 158°, he flies for a further 30 km.

    a Make an accurate drawing of his route.

    b At the end, how far is he away from his starting point?

3 Mike buys a ticket to fly from London to New York for £349·99.
    There is a £55 extra charge to fly from his local airport to London.
    How much does it cost him altogether?

4 Sue's take-home pay for April is £963·20.
    Her total deductions are £374·05.
    Calculate her gross pay for April.

5 Mr Yates sells central heating systems.
    He is paid commission of 4% on each sale.
    How much commission does he get on a system which sells for £2000?

6 Copy and complete these distance tables.

    a Average speed is 25 m/s

Time	1 s	4 s	6 s	9 s	12 s
Distance					

    b Average speed is 8 m/s

Time	5 s	7 s	9 s	17 s	30 s
Distance					

7 a Calculate the area of:
      i the front of the box
      ii the side of the box
      iii the top of the box.

    b What is the total area of the six faces?

    c Calculate the volume of the box.

(Box diagram: 30 cm height, 20 cm width, 12 cm depth — "Wash It")

8 Theresa and Tom hire a taxi.
    After 1 mile, the meter reads £3.
    After 2 miles, the meter reads £5.
    After 3 miles, the meter reads £7.

    a Copy and complete the table.

No. of miles	1	2	3	4	5	〉〈	15
Meter reading (£)	3	5	7				

    b What is the rule for calculating the cost of the journey from the number of miles?

REVISE

9 Calculate the size of the missing angle.

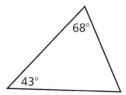

10 Stewart is learning to draw cartoons.
He has been given a basic exercise to draw a face with eyes,
nose and mouth.
The eyes may be open or closed. The nose can be a point or a circle.
The mouth can be a smile or a frown.

This picture shows: eyes closed, circle nose, smile.
Make a table like this, listing all the possible pictures.

Eyes	Nose	Mouth

11 Connor noted the costs of magazines in a rack:
£2·99, £2·50, £3·50, £2·00, £4·99, £1·45, £3·50, £3·45, £3·99.
Calculate the mean cost of a magazine in the rack to the nearest penny.

12 The graph shows how the sales of
gloves change as time passes.

a How many pairs were sold
  i 1 week after January 1
  ii 4 weeks after
  iii 8 weeks after?

b Describe the trend in glove sales.

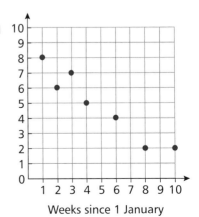

Test F Part 1

1 Calculate:

  a 5403 − 787   **b** 8295 × 9   **c** 4074 ÷ 7   **d** 90·1 ÷ 100

2 a Round 787 to the nearest 100.

  b Round 14·5 to the nearest whole number.

3 Calculate $\frac{1}{5}$ of 97·65.

4 What is $33\frac{1}{3}\%$ of £564?

5 This is a plan of the town near John's house.

  a Give directions for John to go from his house to the bank.

  b Simon turned right out of the bus station, took the first left then first right.
He then crossed over the next road he came to. What was on his right?

6 This is Mr Bridge's weekly payslip.

Name T. Bridge	Employee number 32	NI number YM743275A	Week number 23
Basic pay £350	Overtime £40	Bonus £30	Gross pay
Income tax £65	NI £45	Pension £20	Total deductions
			Take-home pay

Calculate his **a** gross pay   **b** total deductions   **c** take-home pay

7 Allan looks up the times of golf lessons.

Golf lessons	
Monday	6.40 pm–7.00 pm
Tuesday	6.40 pm–7.10 pm
Wednesday	7.10 pm–8.35 pm
Thursday	8.50 pm–10.20 pm
Friday	7.20 pm–9.25 pm

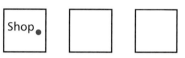

Allan's flow diagram

Start

Is the session less than 1·5 hours?

NO — Attend   YES — Don't attend

Stop

  a How long does the Wednesday session last?

  b Will Allan be attending?

  c Which nights will Allan be missing?

REVISE

8 a Write down the coordinates of
    i P   **ii** Q   **iii** R.

  b P, Q and R are three corners of a rhombus.
    Write down the coordinates of the fourth corner.

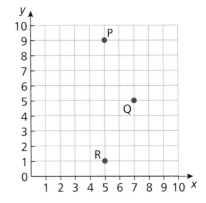

9 Copy this design and draw
the axes of symmetry.

10 Building bricks are being packed into a box.
The bricks are cubes 6 cm long.
The box is 48 cm long, 42 cm wide and 60 cm high.

  a How many bricks would fit along the length of the box?

  b How many bricks would fit along the width of the box?

  c How many bricks would fit into the bottom of the box?

  d How many layers of bricks would fit into the box?

  e How many bricks in total would fit into the box?

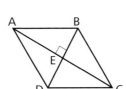

11 ABCD is a rhombus.
BE = 4 cm and CE = 7 cm.

  a Calculate the area of ΔBEC.

  b Calculate the area of rhombus ABCD.

12 Children were given 10 words to spell in a competition.
Here are their scores:

8	1	5	1	4	7	6	6	4	10
8	6	5	8	2	1	3	8	10	4
7	8	1	8	5	3	7	10	10	7

  a Copy and complete the frequency table.

Score	1	2	3	4	5	6	7	8	9	10
Tally										
Frequency										

  b What was the modal score?

  c How many children scored above this mark?

  d How many took part in the competition?

REVISE

Test F Part 2

1 Calculate:

   a $24 \cdot 02 - 4 \cdot 1$ **b** $21 \cdot 29 \times 7$ **c** $16 \cdot 29 \div 9$

2 The front faces of two speakers are similar rectangles.

   a What is the enlargement scale factor?

   b Calculate the missing height.

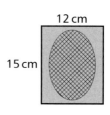
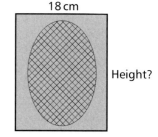

3 Calculate:

   a the VAT

   b the total cost of the bill.

Electricity bill	
Cost of electricity	£64·40
VAT at 5%	_____
Total bill	_____

4 Ruth's hourly pay is £6.
She works 8 hours each day from Monday to Friday.
How much does she earn:

   a for an 8 hour day   **b** in 1 week   **c** in a year (52 weeks)?

5 a   Distance = (speed in km/h) × (time taken in hours)

   Use this formula to help you find the distance for a journey that takes:
   i 3 hours at a speed of 20 km/h   **ii** $\frac{1}{2}$ hour at a speed of 10 km/h

   b   Speed in mph = (distance in miles) ÷ (time taken in hours)

   Use this formula to find the speed on a journey of:
   i 100 miles taking 5 hours   **ii** 48 miles taking 4 hours

6 Add four more tiles of each shade to the grid.

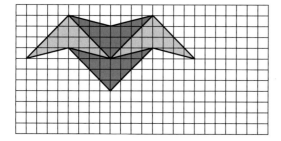

7 Zara sails 30 km from port on a bearing of 100°.
She then turns on a bearing of 200° and sails for a further 40 km.

  a Make an accurate drawing of her voyage.

  b At the end, how far is she away from port?

8 Samuel flies from London to Vancouver for £874·94.
He pays an extra 5% because the plane must land in Seattle first.
How much does it cost him altogether?

9 Freda's basic pay for June is £963·20.
She is given a summer bonus of £350.
Her total deductions are £374·05.
Calculate her gross pay for June.

10 A postman weighed the parcels he put in the sack:
2·4 kg, 1·2 kg, 1·8 kg, 2·1 kg, 2·7 kg, 1·6 kg, 0·9 kg, 2·6 kg

  a What is the total weight?

  b What is the mean weight of a parcel in his sack?

11 A teacher charted the absentees in his class over a week.

  a How many absentees were there on Friday?

  b How many more were there on Wednesday than on Monday?

  c What is the general trend showing up on the graph?

  d Calculate the mean absentees per day.

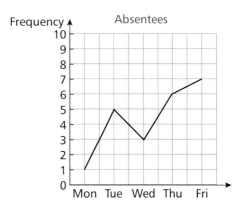

12 This is Jenny Murray's weekly payslip.

Name	Employee number	NI number	Week number
J. Murray	712	YT7821354D	50
Basic pay	**Overtime**	**Bonus**	**Gross pay**
£357·25	£75·60	£45	
Income tax	**NI**	**Pension**	**Total deductions**
£78·56	£36·24	£42·25	
			Take-home pay

Calculate her:

a gross pay

b total deductions

c take-home pay.